THE DARWINIAN PARADIGM

THE DARWINIAN PARADIGM

Essays on its history, philosophy, and religious implications

MICHAEL RUSE

ROUTLEDGE
London and New York

First published 1989

First published in paperback 1993
by Routledge
11 New Fetter Lane, London EC4P 4EE

Simultaneously published in the USA and Canada
by Routledge
29 West 35th Street, New York, NY 10001

© 1989, 1993 Michael Ruse

Printed in Great Britain by
T.J. Press (Padstow) Ltd., Padstow, Cornwall

British Library Cataloguing in Publication Data
Ruse, Michael
The Darwinian Paradigm: essays on its
history, philosophy, and religious
implications.
1. Organisms. Evolution. Philosophical
perspectives
I. Title
116
.Library of Congress Cataloging in Publication Data
Ruse, Michael.
The Darwinian Paradigm: essays on its history, philosophy, and
religious implications / by Michael Ruse
p. cm.
Bibliography: p.
Includes index.
1. Evolution. 2. Evolution–Philosophy. 3. Natural selection.
4. Darwin, Charles, 1809-1882. I. Title.
QH371.R8 1989
575-dc19   88-23985
ISBN 0-415-08951-4

For William and Ursula Ruse with affection

Aristotle's *Physica*, Ptolemy's *Almagest*, Newton's *Principia* and *Opticks*, Franklin's *Electricity*, Lavoisier's *Chemistry*, and Lyell's *Geology*–these and many other works served for a time implicitly to define the legitimate problems and methods of a research field for succeeding generations of practitioners. They were able to do so because they shared two essential characteristics. Their achievement was sufficiently unprecedented to attract an enduring group of adherents away from competing modes of scientific activity. Simultaneously, it was sufficiently open-ended to leave all sorts of problems for the redefined group of practitioners to resolve.

Achievements that share these two characteristics I shall henceforth refer to as 'paradigms'.

Thomas Kuhn,
The Structure of Scientific Revolutions (1962)

CONTENTS

ACKNOWLEDGEMENTS

The following essays have been published before. I am grateful for permission to include them in this collection.

'Darwin's debt to philosophy: an examination of the influence of the philosophical ideas of John F.W. Herschel and William Whewell on the development of Charles Darwin's theory of evolution', *Studies in History and Philosophy of Science* 6 (1975) 159-81.

'Charles Darwin and group selection', *Annals of Science* 37 (1980) 615-30.

'What kind of revolution occurred in geology?' in P. Asquith and I. Hacking (eds) *PSA 1978*, East Lansing, MI: Philosophy of Science Association, 1981, 2: 240-73.

'Biological species: natural kinds, individuals, or what?' *British Journal for the Philosophy of Science* 38 (1987) 225-42.

'Teleology and the biological sciences', in N. Rescher, (ed.) *Current Issues in Teleology*, Landham, MD: University Press of America, 1986, 56-64.

'Biological science and feminist values', in P. Asquith and P. Kitcher (eds) *PSA 1984*, East Lansing, MI: Philosophy of Science Association, 1985, 2: 525-42.

'Is rape wrong on Andromeda? An introduction to extraterrestrial evolution, science, and morality', in E. Regis, (ed.) *Extraterrestrials: Science and Alien Intelligence*, Cambridge: Cambridge University Press, 1985, 43-78.

INTRODUCTION

I first read Charles Darwin's masterpiece, *On the Origin of Species*, some twenty years ago. At once I fell under its spell – an emotion which is as strong within me today as it was then. For all of Darwin's problems and gaps and inconsistencies (and over the years I have played my part in bringing these to light) at base, I think he was right. With Darwin, I believe that the organic world came into being through a natural process of evolution, that by far the main mechanism was natural selection, something which speaks directly to the most pervasive feature of organisms namely their adaptedness, and that (although this is only hinted at in the *Origin*) for all of the obvious qualifications one must make, the ideas apply absolutely and completely to humans.

In various writings, I have explored aspects of 'Darwinism', most fully in what evolved into a trilogy of works: on history, *The Darwinian Revolution* (1979); on science, *Darwinism Defended* (1982); and on philosophy, *Taking Darwin Seriously* (1986). And yet still I feel dissatisfied – or rather, still I feel my understanding is just beginning. One thing in particular which puzzles me more and more as the years go by is why so many of my fellow professionals, particularly the philosophers of science, tend not to have the same overwhelming convictions about Darwinism as I. One must not exaggerate. I do not pretend to be a better or more profound scholar than anyone else, nor am I (explicitly or implicitly) implying that most people do not take seriously evolution or natural selection or our own natural origins. I am not even saying that I am the truest or most orthodox Darwinian that there is. In biology itself, there are many more ardent pretenders to that title.

Nevertheless, it is true that a large number of people feel that they should revise or supplement Darwin's thinking about mechanisms with various alternatives and additions, while at the same time they resist the enormous commitment to organic adaptedness which so pervades the *Origin*. And when it comes to humankind and today's extensions of Darwinism, there is a positive philosophical stampede to other positions. What I look upon as thrilling moves forward, others regard as pernicious collapses into darkness and confusion.

The obvious reason for all of this, a reason which has tempted me more than once, is that I am right and others are wrong. An even more satisfying reason is that, when it comes to our own species, I have the courage of my convictions and others do not. But I am coming to see that matters are more complex and interesting than this. What separates the ultra-Darwinian like myself from the critics and doubters and revisers and extenders is less a simple question of fact and argument, and more one of general perspective. As I have said: at base, I think Darwin was right. Others do not.

This all leads one to suspect, subject to qualifications and reservations, that we have here what Thomas Kuhn in his *Structure of Scientific Revolutions* described as a 'paradigm' difference: a gap between different world pictures. This, at least, without wanting to impose an artificial sense of unity, is the theme I hope to illustrate in this collection. I want to show you just why Darwinism, even (especially) extended to humans, just 'feels right' to me. At the same time, I hope I shall avoid being mushy and mystical. Evolution through natural selection must succeed on its own merits, and not through ill-defined yearnings for meaning.

My aim is primarily positive rather than negative. With some few exceptions, I am much more interested in defining and expanding my vision of Darwinism than in criticizing others. Basically, what I hope to show is how one thinker, over the past several years, has taken the legacy of the *Origin* and tried to understand himself and the world around him. For me, certainly, Charles Darwin's achievements have had the two essential characteristics of paradigmhood.

I am happy to acknowledge that over the years I have received much help and advice from many historians, philosophers, biologists, and most recently (especially through the Institute on Religion in an Age of Science) theologians. Closer to home, in preparing this collection I have been helped by my research assistant, Constance Matthews-Cull and my secretarial assistant, Gail McGinnis. I have left unchanged previously published essays; they must stand on their own, warts and all. However I have imposed a uniform style and collected all references into one joint bibliography.

Part I

HISTORICAL THEMES

These first three essays look toward the past, but, perhaps uncomfortably so for most of today's historians of science, they were written with at least one and a half eyes on our thinking today. The first on Darwin and the philosophers, the oldest in the collection, explores the structure of Darwin's thinking in the *Origin* and some of the influences on this thought. I was certainly not the first to pick up on this aspect of Darwin's theory – pioneering work was done previously by Michael Ghiselin (1969) and David Hull (1973a) – but I would like to think I carried debate forward and helped provide a foundation for what is now a cottage industry.

In a funny way, however, I now see the real strength of the paper in something for which it was not primarily intended. Then, my real aim was to further and support the logical empiricist philosophy of science, a viewpoint which stresses that scientific theories consist of laws of nature bound together in tight deductive (axiomatic) structures – the best exemplar of this being Newtonian mechanics. Having analysed contemporary biology from this perspective (Ruse 1973c), I wanted then to show that Darwin's work fits the pattern (see also Ruse 1975c). I still think there is much life in logical empiricism (Ruse 1981c), but for me what counts now about this paper is the way it shows that Darwin took seriously the leading methodologists of his day. Even though Darwin has a rather easy, self-depreciating style, you should not think his ideas rest simply on the surface. The theory of the *Origin* is a very subtle piece of work. Darwin's sheer professionalism, reflected in his theory, is what counts.

The second essay of the section backs up this point. I certainly would not pretend that every idea we hold dear today was back there in 1859, the year in which the *Origin* was published. It seems to me that Darwin was hopelessly confused about heredity, and no amount of special pleading will prove otherwise. Yet, despite his notorious inability to think mathematically, Darwin's thought was often sufficiently sophisticated to bear re-examination for insights on problems which plague us today. One such worry is that focusing on the level at which natural selection is supposed to operate. Crucial to modern Darwinism, both in its application generally (especially over long periods of time) and in its application specifically to humankind, is the belief that selection works almost exclusively on the individual. There is no place for selection of collections of organisms, whether this be through so-called 'group selection' or through so-called 'species selection'. (See Brandon and Burian 1984 for details.) As we shall see later, this stance has major implications for our thinking about social behaviour. And as we see here, although about twenty years ago evolutionists with great fanfare discovered the merits of individual selection, Darwin was before them. He had already thrashed out the pertinent issues with natural selection's co-discoverer Alfred Russel Wallace, and had taken a firm individualistic stance for the very reasons which guide today's thinkers.

Finally, in the one essay in the whole collection not directly on biology, I consider the nature of the recent revolution in geology. I include this essay for three reasons. First, because historically Darwin started in science as a geologist, and moreover was hoping to find *the* overarching causal theory, as he was later to succeed in doing in biology. The arrival and acceptance of plate tectonics was the successful culmination of Darwin's own programme (although he himself thought of continents more in terms of their moving up and down than sideways). Second, because, as I argue, the influences on today's geologists are precisely those which were on Darwin, and they succeeded for the same sorts of methodological reasons as he. Third – and this starts to point us towards themes to be considered in the next section – because in the essay I explore elements of Kuhn's thinking, particularly about the nature of paradigms. This prepares ground for thoughts which I have about the paradigmatic nature of Darwinism and its rivals.

Parenthetically, let me note that when I wrote this third essay about ten years ago, there was almost no philosophical analysis of the geological revolution. That struck me as a scandal then and still so strikes me. However, I must note that, apart from the symposium in which this essay first appeared, seminal studies of the episode have been produced by Henry Frankel. (See especially Frankel 1979.)

Chapter One

DARWIN'S DEBT TO PHILOSOPHY

Charles Darwin went up to Cambridge as an undergraduate in 1828.[1] He set off on his voyage around the world on the *Beagle* in 1831, returning in 1836. About the time of his return he became an evolutionist, and he hit upon the evolutionary mechanism for which he is most famous, natural selection brought on by the struggle for existence, in the autumn of 1838. In 1842 he wrote a short sketch of his theory, and in 1844 he expanded this into a fairly substantial essay (Darwin and Wallace 1958). At the urging of his friends, in 1856 he started to prepare for publication a massive evolutionary work incorporating his basic ideas (Stauffer 1959, Darwin 1975). This work was interrupted by the arrival of A.R. Wallace's essay on evolution, one in which he mirrored Darwin's ideas in an uncanny fashion, in 1858. Thereupon, Darwin dropped all else, wrote an 'abstract' of his evolutionary ideas, and this was published as the *Origin of Species* in 1859.

In this paper I argue that an important factor in Charles Darwin's development of his theory of evolution through natural selection was the philosophy of science in England in the 1830s. When this factor is recognized, then new light is thrown both upon Darwin's discovery of his evolutionary mechanism and upon the way in which he prepared his theory for public presentation.

PHILOSOPHY OF SCIENCE, 1830–40

England's most influential philosopher of science in the 1830s was the famed astronomer, John F.W. Herschel, whose philosophical reputation rested upon the deservedly popular *Preliminary*

Discourse on the Study of Natural Philosophy (1831).[2] Not surprisingly, for Herschel the paradigmatic sciences were the physical sciences, particularly Newtonian astronomy (of the 1830s), and the claims Herschel made about the way science is, or ought to be, reflect this bias. Consequently, many of Herschel's major claims have a curiously familiar ring to today's reader, for in important respects he anticipated the modern philosophical school which also looks to physics for its ideals, so-called 'logical empiricism'. I shall now sketch those tenets of Herschel's philosophy which might have been of interest to a budding scientist; that is, I shall ignore Herschel's metaphysical speculations on the ultimate nature of science and concentrate exclusively on his methodology. I shall consider what, in Herschel's opinion, was the kind of theory a scientist ought to aim for and the kind of evidence a scientist ought to offer. I shall not at present consider any methodological directives that Herschel thought peculiarly applicable to the biologist, although I disclose no secrets by admitting that Herschel was not sympathetic towards evolutionary theories.

Essentially Herschel saw scientific theories as hypothetico-deductive systems. Thus he wrote that

> the whole of natural philosophy consists entirely of a series of inductive generalizations ... carried up to universal laws, or axioms, which comprehend in their statements every subordinate degree of generality, and of a corresponding series of inverted reasoning from generals to particulars, by which these axioms are traced back into their remotest consequences, and all particular propositions deduced from them. (Herschel 1931, p. 104)[3]

Moreover, Herschel made clear that what distinguishes scientific axiom systems from other such systems is that the former, unlike the latter, contain laws; these are universal, empirical statements 'of what will happen in certain general contingencies' (p. 98). What elevates a law above a mere catalogue of empirical facts is that in some sense it expresses the way things must be, that is, to use modern terminology, it allows for 'counterfactuals': if *A* were to occur (even if it does not), then *B* would follow. 'Every law is a provision for cases which *may* occur, and has relation to an infinite number of cases that never have occurred, and never will' (p. 36).

Herschel distinguished upper level laws, 'fundamental laws,' from lower level (derived) laws, or 'empirical laws' (1831, pp. 178, 200). Newton's laws of motion and gravitation are the highest of all fundamental laws, Kepler's laws are prime examples of empirical laws (p. 178). It goes almost without saying that although empirical laws have an indispensible role in science, the ultimate aim of the scientist is fundamental laws, and there are strong hints in Herschel of the distinction modern logical empiricists draw between observable and unobservable concepts (reference to the latter occurring in the axioms of a scientific system and reference to the former occurring in the lower-level derived laws of the system). Thus Herschel wrote that 'the agents employed by nature to act on material structures are invisible, and only to be traced by the effects they produce' (p. 193). Herschel argued also that the best kind of fundamental or higher law is *quantitative*; for instance, the law of gravitation, 'the most universal truth at which human reason has yet arrived' (p. 123), gives exact ratios for gravitational attractions.

One point which Herschel emphasized at length is the need of the scientist to make reference in his fundamental laws to (and thus to explain through) *causes*. In particular, the scientist should aim at explaining through *verae causae*, where these are causes 'competent, under different modifications, to the production of a great multitude of effects, besides those which originally led to a knowledge of them' (1831, p. 144). In other words, the scientist must aim to get away from *ad hoc* putative causes, proposed just to explain one set of phenomena; he must try to relate phenomena of different kinds and to explain them through one embracing all-sufficient cause or mechanism. Only then can the scientist be reasonably certain that he has 'causes recognized as having a real existence in nature, and not being mere hypotheses or figments of the mind' (p. 144). Needless to say, at the top of *verae causae* is force; indeed, Herschel speculated whether all causes might not reduce ultimately to some kind of force (p. 88).

Finally, what should be mentioned is a point Herschel made so frequently about the confirmation of theories that it might well be regarded as the *leitmotif* of his book, namely that the mark of a truly confirmed theory, one which absolutely has to be taken as true and resting on a *vera causa*, is that the theory be found to explain phenomena in ways unanticipated when the theory was first devised or to explain phenomena which seemed hostile to the theory when first devised.

> The surest and best characteristic of a well-founded and extensive induction, however, is when verifications of it spring up, as it were, spontaneously, into notice, from quarters where they might be least expected, or even among instances of that very kind which were at first considered hostile to them. Evidence of this kind is irresistible, and compels assent with a weight which scarcely any other possesses. (Herschel 1831, p. 170; see also pp. 29-34, 97-8)

The other important philosopher of science in the period being considered was Herschel's close friend, William Whewell.[4] Herschel and Whewell came to differ quite considerably over what I have called the 'metaphysical' aspects of science, Herschel inclining more to empiricism whereas Whewell was much influenced by Kant. However, they differed little, if at all, with respect to 'methodological' questions, the kind of theory a scientist should aim to produce and the way he should try to confirm it. This is perhaps not surprising because, I think, Herschel and Whewell worked out their philosophies far more in conjunction than independently, and (the Cambridge-educated) Whewell agreed fully with (the Cambridge-educated) Herschel that the finest of all sciences is Newtonian mechanics, particularly Newtonian astronomy. Indeed, in an address to the British Association in 1833 Whewell spoke of Newtonian astronomy as being the 'queen of the sciences',[5] and in his *History of the Inductive Sciences* he wrote that

> Newton's theory is the circle of generalization which includes all the others; the highest point of the inductive ascent; the catastrophe of the philosophic drama to which Plato had prologized; the point to which men's minds had been journeying for two thousand years. (Whewell 1837, 2, p. 183)

Whewell's major work on the philosophy of science, *The Philosophy of the Inductive Sciences*, did not appear until 1840; but in various writings in the 1830s he managed to show his support of many of the important tenets of Herschel's philosophy. Thus, for instance, Whewell wrote an enthusiastic review of Herschel's *Discourse* in the *Quarterly Review* for April 1831. He adopted and emphasized Herschel's point about the best kind of laws being quantitative laws. Then in 1833, in his book on natural theology, Whewell agreed not only that the aim of science is to find laws, 'rules describing the mode in which things *do* act; [things] invariably obeyed' (Whewell 1833b), but he advocated, explicitly, the hypothetico-deductive ideal for science (p. 325). And then in his *History*, Whewell followed Herschel in distinguishing between two kinds of laws, speaking of 'formal' or 'phenomenal' laws and 'physical' or 'causal' laws, the models for this division being, once again, Kepler and Newton (Whewell 1837, books 5 and 7).

Finally, there is the question of confirmation. In his *History* Whewell was at great pains to show that the strength of great theories, particularly Newtonian mechanics, is the ability to explain in many different areas, including those unthought of before the theory was discovered.[6] As is well known, in his *Philosophy* Whewell labelled this process the 'consilience of inductions', and, like Herschel, made much of the element of surprise: 'the evidence in favour of our induction is of a much higher and more forcible character when it enables us to explain and determine cases of a *kind different* from those which were contemplated in the formation of our hypothesis' (Whewell 1840, 2, p. 230). Hence, both with respect to theory-nature and with respect to theory-proof Herschel and Whewell spoke with almost one voice.[7]

DARWIN AND THE PHILOSOPHERS

That Darwin was aware of and responded positively to this philosophy of science is undeniable. Take the influence of Herschel. Darwin first read Herschel's *Discourse* early in 1831; he reacted enthusiastically to it at the time, urging his cousin to 'read it directly',[8] and, late in life looking back over his career, he spoke of Herschel's work in the highest possible terms.

> During my last year at Cambridge I read with care and profound interest Humboldt's Personal Narrative. This

work and Sir J. Herschel's *Introduction to the Study of Natural Philosophy* stirred up in me a burning zeal to add even the most humble contribution to the noble structure of Natural Science. No one or a dozen other books influenced me nearly so much as these two. (Darwin 1969, pp. 67-8)

Darwin reread the *Discourse* late in 1838,[9] by which time he knew Herschel personally. Their social circles overlapped and, more interesting, they both appear to have been active members of the (London) Geological Society.[10] Darwin wrote of Herschel that 'He never talked much, but every word which he uttered was worth listening to' (1969, p. 107). I shall show later that Darwin always thought highly of Herschel and craved his praise.

Darwin's relationship with Whewell is most interesting. Whewell was a violent anti-evolutionist, and I suspect that in later life neither he nor Darwin was over-keen to emphasize their earlier intimacy. But such intimacy there certainly was. Whilst an undergraduate Darwin knew Whewell well: for his full three years at Cambridge Darwin attended the lectures on botany by the Revd. J.S. Henslow, as also did Whewell.[11] Whewell and Darwin met also at Henslow's weekly scientific evenings, Darwin walking home with Whewell. About Whewell Darwin wrote that 'Next to Sir J. Mackintosh he was the best converser on grave subjects to whom I ever listened' (1969, p. 66). It goes without saying that, given the context, these 'grave subjects' would have included much about science: no doubt in 1831 the enthusiastic Darwin and the equally enthusiastic Whewell talked about Herschel's *Discourse.*

After his return from the *Beagle* voyage Darwin lived (early in 1837) in Cambridge for three months, but his most important contact with Whewell was through the Geological Society. Whewell was president in 1837 and 1838 whilst Darwin was on the council, and this led to fortnightly meetings.[12] Whewell seems to have pushed Darwin's scientific career strongly: he urged him to get on with the publishing of the results of the *Beagle* voyage, he pressed him into accepting a secretaryship of the Society;[13] in his second presidential address to the Society, he heaped the highest possible praise on Darwin (and hinted, incidentally, that he felt some credit due to himself as one of Darwin's teachers). In letters to Whewell, Darwin thanked him for having 'shown so much interest and kindness in all my affairs' and for 'the manner of your whole intercourse with me, since my return to England'.[14]

I think Whewell's major influence on Darwin would have been through conversation, but Darwin did read several things by Whewell. These include Whewell's address to the British Association (Whewell sent Darwin a copy[15]), the *Bridgewater Treatise* (Darwin read this twice, in early 1838 and in 1840[16]), and the *History*. Darwin owned a copy of this last-named work; he skimmed it at some point in 1838, probably in early October, and then, just after his rereading of Herschel, read it very carefully, annotating it fully.[17] He liked the work, praising it to Whewell and to others.[18] Moreover, Darwin who was notoriously so careless of his books, had the volumes leather-bound. I doubt if Darwin ever read Whewell's *Philosophy*, but he did respond with great interest to a large detailed review of Whewell by Herschel. ' – From Herschel's Review Quart. June 41 I see I MUST STUDY Whewell on Philosophy of Science.'[19]

Darwin was therefore fully aware of the Herschel-Whewell philosophy of science, and all the direct evidence points to an enthusiastic reaction. Moreover, the genuineness of this reaction is supported, both by comments which Darwin made about scientific methodology and by the scientific works which he produced. We have seen that central to the philosophy was the taking of Newtonian astronomy as the paradigm for science. Many comments made by Darwin show that he accepted this claim entirely, and that, indeed, his aim was to be the Newton of biology. Thus, for example, he wrote as follows in a private notebook in 1837.

> Astronomers might formerly have said that God ordered each planet to move in its particular destiny. In same manner God orders each animal created with certain form in certain country, but how much more simple and sublime power let attraction act according to certain law, such are inevitable consequences – let animal be created, then by the fixed laws of generation, such will be their successors. Let the powers of transportal be such, and so will be the forms of one country to another. – Let geological changes go at such a rate, so will be the number and distribution of the species!! (Darwin, B, pp. 101-2)

And when he was presenting his theory again and again Darwin defended himself against possible criticisms on the grounds that he was being more Newtonian than any would-be critics. Thus, in his first full-length exposition of his theory (the *Essay* of 1844), Darwin asked 'shall we then say that a pair, or a gravid female, of each of these three species of rhinoceros, were separately created ...? For my own part I could no more admit [this] proposition than I could admit that the planets move in their courses, and that a stone falls to the ground, not through the intervention of the secondary and appointed law of gravity, but from the direct volition of the Creator' (Darwin and Wallace 1958, pp. 250-1).

Were one to single out from the Herschel-Whewell philosophy the two features most likely to be manifested in any scientific theory consciously influenced by the philosophy, they would probably be: first, the hypothetico-deductive model, and secondly the use of one central mechanism or cause to explain phenomena in widely different areas. Both of these features are manifested, to a significant extent, in Darwin's theory in the *Origin*, and they can be traced back to Darwin's earlier versions of his theory, the *Sketch* of 1842 and the *Essay* of 1844.[20] Furthermore, these were features Darwin intended his theory to have and he took pride in the fact that (as he thought) his theory did have them.

Take first the hypothetico-deductive ideal. Darwin's following of this is particularly apparent in what one might call the 'core' arguments of his theory. Darwin's major mechanism of evolutionary change, natural selection, is something which embodies the notion that in each generation there is a differential reproduction of organisms, more organisms being born than can survive and reproduce, and the notion that the survival of the successful organisms is in part a function of characteristics which they, unlike unsuccessful organisms, possess. Darwin did not just drop natural selection into his theory, unannounced. Rather, he argued first to a struggle for existence and then to natural selection, and these arguments to the struggle and then to natural selection approximate closely to the hypothetico-deductive ideal (Ruse 1971). Thus Darwin started his arguments with statements which seem very much like laws (understood in the Herschelian sense), for instance, that given any species of organisms they will be found to have a tendency to increase their numbers at a geometrically high rate. And this, he tried to show, is something which *must* hold for any species you like to name, even the most slow breeding of species. Then, from lawlike statements like these, Darwin tried to show that his conclusions, first about a struggle and then about selection, *must* follow. And, of course, this is what deduction is all about.

Even more obvious than Darwin's attempt to satisfy the hypothetico-deductive ideal was his attempt to use his mechanism of evolutionary change, natural selection, to explain phenomena in many widely different areas. Thus Darwin showed how natural selection solves problems of geographical distribution, of instinct, of geology, of classification, of comparative anatomy, of embryology, and so on. All of these various areas come under the umbrella of selection just as so many areas of physical enquiry come under the umbrella of Newtonian gravitational force. And, as I have mentioned, Darwin intended and took credit for having shown both this fact and the former fact, namely that he had manifested the hypothetico-deductive ideal. He wrote constantly of showing how things, first like the struggle and then like the phenomena of geographical distribution, follow 'inevitably' from laws. (See Darwin 1859, pp. 80-1, 489-90.) And whenever challenged about the truth of this theory Darwin pointed always to the wide scope of his mechanism: 'I must freely confess, the difficulties and objections are terrific; but I cannot believe that a false theory would explain, as it seems to me it does explain, so many classes of facts' (Darwin and Seward 1903, 1, p. 455).

It cannot be denied, as critics were quick to point out, that Darwin was not entirely successful at achieving the Herschel-Whewell theory ideal. (See Hopkins 1860.) In particular, many of the inferences in Darwin's theory taken as a whole were far from being rigorously deductive. However, this is not to deny Darwin's intentions, and one's estimation of the success he actually achieved becomes much increased when one compares Darwin's theory against the works in the 1830s of other non-physical scientists. Thus, although Lyell's (1830-3) chief aim was to show that the past world can be explained by laws of the present world, he never achieved even the limited hypothetico-deductive success of Darwin, preferring rather to make his points with strings of related examples. And the same goes for the work of someone like Henslow (1835), who relied on description and example rather than the axiomatic method.

In concluding this section, let me make one caveat. I argue that Darwin was influenced by the Herschel-Whewell theory ideal and I have given reasons to suggest that this would have been a direct influence. I do not, however, want to suggest that this was an entirely exclusive influence. I think that pretty well everybody in the 1830s accepted this philosophy of science and that Darwin would have received it from others as well. For example, Lyell and Whewell had a continuing debate over whether one ought to be a uniformitarian or catastrophist in geology, and both the uniformitarian Lyell and the catastrophist Whewell defended their respective positions as being more Newtonian than the other's![21] And I am sure that a major reason why Darwin did not change his theory in any significant way after its first formulation was because there was no significant change in the philosophy of science (*qua* theory-nature ideal) between the writing of the *Sketch* (1842) and the writing of the *Origin* (1858-9). Even J.S. Mill, in his influential *System of Logic* (1843), managed to incorporate many of the salient features of the hypothetico-deductive approach, though he differed from Whewell at least in his estimate of the sufficiency of that method to yield a doctrine of proof. But then, as I shall show later, at this point where Mill diverged from Herschel and Whewell, Darwin sided with the earlier philosophers rather than with Mill.

But, whilst admitting this caveat about other possible influences on Darwin one must be careful not to underestimate Herschel and Whewell themselves, and certainly one must be careful not to fall into the trap of thinking that because Herschel and Whewell were anti-evolutionists they cannot have been significant influences on Darwin. Nigh-on everyone was an anti-evolutionist in the 1830s; Lyell, probably Darwin's greatest intellectual influence, was one of the leaders of the attack against evolutionary theories, and indeed, Lyell's position was practically indistinguishable from Herschel's.[22] Nor should one assume that Darwin's theory was bound to be the way it was, because every scientific theory was that way. As I have just pointed out, Darwin hardly got the salient aspects of the Herschel-Whewell philosophy from the work of men like Lyell and Henslow, because these aspects were absent from their work. The direct influence of Herschel and Whewell, although not exclusive, should not be discounted; in any case, many of Darwin's other influences like Lyell and Henslow probably got their philosophy of science from Herschel and Whewell in the first place.

NEW LIGHT ON DARWIN

I shall argue now that recognizing the importance for Darwin of the Herschel-Whewell philosophy of science enables us to solve several puzzles in the Darwinian story. I take first the question of Darwin's discovery of his theory, and in particular the role played in this discovery by the thought of Malthus. As mentioned earlier, we know that Darwin came upon, or recognized, his main mechanism of evolutionary change some time in the autumn of 1838. His discovery seems to have been a two-part process; he grasped the principle of natural selection by analogy from breeders' use of artificial selection on domestic organisms, and then, after reading the *Principle of Population* by T.R. Malthus, he saw in some way how he could use the struggle for existence as a kind of driving force behind natural selection. Thus, to Wallace, Darwin wrote: 'I came to the conclusion that selection was the principle of change from the study of domesticated productions; and then, reading Malthus, I saw at once how to apply this principle' (Darwin and Seward 1903, 1: 118).

Recent Darwin scholarship has shown that Darwin's route to discovery was less direct than he himself implied. (See Herbert 1971, Limoges 1970.) For a start, before the reading of Malthus (about 28 September 1838) most of the comments Darwin made show that he, like everyone else at the time, looked on the domestic world as pointing *away* from a mechanism of evolutionary change, rather than towards it. For instance, one comment Darwin made shortly before reading Malthus was: 'It certainly appears in domesticated animals that the amount of variation is soon reached – as in pidgeons no new races' (notebook D, p. 104, written 13 September 1838). However, despite comments like these, it does now seem that Darwin was definitely led to the mechanism of natural selection from the analogy with artificial selection. In particular, Darwin got the concept of natural selection in mid-1838 from reading animal breeders' pamphlets, which pamphlets talked not only about artificial selection but also about natural selection (not by that name); and explicitly drew an analogy between the two kinds of selection.[23] Nevertheless, a puzzle about Malthus still remains. Why was it necessary for Darwin to read Malthus before he recognized that in natural selection he had a mechanism of evolutionary change? Before reading him Darwin gave no hint that he differed from the breeder's assessment of natural selection, namely that it was something which would cause only limited change *within* a species. It cannot be just that Malthus drew Darwin's attention to the struggle for existence, because Darwin knew all about the struggle long before reading him. The struggle is described explicitly and in detail in Lyell's *Principles of Geology*, two editions of which Darwin had read by mid-1837. Indeed, Lyell even talks of the struggle for existence *by that name*.[24]

Understanding the importance for Darwin of the Herschel-Whewell philosophy, Malthus' contribution to Darwin's discovery becomes readily explicable. Malthus showed Darwin how he could locate the struggle, with the consequent selection, in a hypothetico-deductively organized network of laws; of laws which were, moreover, *quantitative*: in Herschel's and Whewell's eyes the best kind of laws. Malthus argued that a struggle for existence amongst humans would inevitably ensue, unless prevented by moral restraint (or something unmentionable like contraception), because humans have a tendency to increase in number at a *geometrical* rate whereas their food supplies can increase only at maximum at an *arithmetical* rate. Darwin seized upon this argument, generalizing to all animals, thus eliminating the alternatives to the struggle. (See Ruse 1973b.) He now had quantitative laws, leading deductively to the struggle, which he was then able to extend to selection. Thanks to Malthus, Darwin was able to put his mechanism for evolutionary change into a satisfactory context, a context, that is, which satisfied the Herschel-Whewell theory ideal.

But Malthus was important for Darwin for another, related reason. The Herschel-Whewell philosophy demanded that one explain through causes, the best kind of which, perhaps the only kind of which, were forces. Through Malthus, Darwin saw the struggle as being a kind of force, which would in turn, as it were, propel the force of selection. As soon as he read Malthus the excited Darwin scribbled in his notebook that

> Population is increased at geometrical ratio in FAR SHORTER time than 25 years – yet until the one sentence of Malthus no one clearly perceived the great check amongst men. – there is spring, like food used for other purposes as wheat for making brandy. – Even a *few* years plenty, makes population in man increase and an *ordinary* crop causes a dearth. take Europe on an average every species must have same number killed year with year by hawks, by cold etc. – even one species of hawk decreasing in number must affect instantaneously all the rest. – The final cause of all this wedging, must be to sort out proper structure, and adapt it to changes. – to do that for form, which Malthus shows is the final effect (by means however of volition) of this populousness on the energy of man. One may say there is a force like a hundred thousand wedges trying [to] force every kind of adapted structure into the gaps in the oeconomy of nature – or rather forming gaps by thrusting out weaker ones. (Darwin, D, p. 135)

As this passage shows, Malthus enabled Darwin to see the struggle and the consequent selection in terms of force. Hence Darwin, working in the light of the Herschel-Whewell philosophy, felt able to regard selection as a possible evolutionary mechanism.

If, as I argue, the Herschel-Whewell philosophy was an important factor in Darwin's response to Malthus, one might naturally ask if the philosophy played any role in Wallace's discovery of natural selection, because he like Darwin acknowledged an important debt to Malthus (Wallace 1905, 1, pp. 361-2). Although Wallace certainly read Whewell's *History* (McKinney 1972, p. 24), I suspect the real key to Wallace's response lies in Robert Chambers' *Vestiges of the Natural History of Creation*. McKinney argues that 'the influence of the *Vestiges* [on Wallace] ... can scarcely be overemphasized' (1972, p. 12). But a major aim of *Vestiges* is to show that as good Newtonians we must accept a biological evolutionary theory. Wallace, I think, whilst rejecting as inadequate Chambers' own evolutionary theory, entirely accepted Chambers' research programme, to find the biological analogue of Newtonian astronomy.[25] Thus I would suggest that

21

Wallace, like Darwin, may have reacted favourably to Malthus' ideas because he could then start to see his way towards a biological equivalent of Newtonian astronomy. Hence I think that Darwin and Wallace quite possibly started from similar philosophical positions, although I have no reason to believe that they drew on exactly the same immediate sources for the philosophies. Indeed, I doubt that their sources were exactly the same, for, as I shall show next, Darwin took an altogether different methodological step from Wallace because, I think, he wanted to present a theory which would satisfy Herschel's criteria of theory-excellence.

As 1838 drew to a close, Darwin had his major mechanism of evolutionary change. He had now to start to think about converting his mechanism into a full-blown theory, one which he would present to the world. An understanding of the Herschel-Whewell influence remains crucial to the grasping of Darwin's reasonings, particularly the way in which he used the analogy from artificial selection.

Darwin knew well that any theory of evolution was going to be highly controversial, to say the least. That meant he had to make the best possible case, particularly the best possible case in the eyes of the ultimate arbiters of scientific acceptability, Herschel and Whewell. He felt he had to satisfy their criteria of good science. Indeed, interestingly, Darwin always felt this way. By 1859, the year of publication of the *Origin*, the long-invalid Darwin moved in different circles from the philosophers, Whewell particularly. Nevertheless, Darwin sent copies of the *Origin* to both Herschel and Whewell, and he prefaced the *Origin* with a quotation by Whewell to the effect that the world works exclusively according to law (as if to point out that he, Darwin, was merely following Whewell's prescriptions),[26] and, most significantly, waited with interest and trepidation for Herschel's evaluation of his theory. When the great man was reputed as having characterized the *Origin* as 'the law of higgledy-piggledy,' Darwin spoke of Herschel's evaluation as 'a great blow and discouragement' (Darwin 1887, 2, p. 241). (Actually, as we shall see, Herschel's verdict was not entirely negative.)

In December of 1838, Darwin turned seriously to the question of how best he ought to develop and present his theory. To this end, he reread Herschel's *Discourse* and went very carefully through Whewell's *History*. Gauging his interest in the latter work from the extent to which he annotated and marked the various sections of his own copy, Whewell's volumes were of particular interest for two reasons. On the one hand, Darwin wanted to see what were the precise merits of a theory like Newton's, what made it so exceptional a theory. On the other hand, he wanted to see what was the strongest possible case that could be made by an anti-evolutionist: Darwin wanted to leave no possible criticism unconsidered. Thus, when Whewell claimed that every evolutionist would be saddled with Lamarckian assumptions about necessary progressive evolutionary tendencies and constant creation of new sparks of life ('monads'), Darwin exclaimed in the margin that 'These are not assumptions, but consequences of my theory, and not all are necessary' (Whewell, 1837, 3, p. 579).

Now, Whewell's major criticism of the evolutionist, one which was to be found in both Cuvier and Lyell, was that present evidence, particularly that of animal and plant breeders, pointed away from rather than towards the creation of new species. Hence, argued Whewell, new species cannot have been created naturally in the past. Darwin realized that if he were to make his case he had to counter this criticism, and as is well known, the way in which he tried to do this was by arguing that Whewell and others were wrong to cite modern breeding techniques and results as evidence *against* evolution. Darwin argued that in fact such techniques and results were evidence *for* evolution (Ruse 1973b). But why did Darwin employ this strategy? We saw that earlier, in 1838, Darwin himself seems to have agreed that the domestic world points away from rather than towards evolution. Why did Darwin not employ the kind of strategy employed by Wallace in his 1858 evolutionary essay, and argue that since the domestic and the wild worlds are so drastically different, one cannot possibly draw any analogies between the two, and that hence the failure to produce new permanent forms in the domestic world does not prove that no such forms can be produced in the wild world?

Part of the reason why Darwin adopted the particular argument that he did stems, no doubt, from the fact that by the end of 1838 he was beginning to doubt the conventional wisdom on animal and plant breeding; he was starting to get evidence that artificially induced changes could be fairly permanent. But this was not the main reason why he suddenly became so keen to *stress* the analogy between domestic and natural selection;[27] why he suddenly swung round completely from his earlier position and wrote in a notebook 'It is a beautiful part of my theory, that domesticated races of organics are made by precisely same means as species – but latter far more perfectly and infinitely slower' (notebook E, p. 71).[28] The answer to Darwin's switch lies in the doctrine of *verae causae*, a doctrine, as we have seen, that was absolutely central to a Herschellian philosophy of science.

Herschel argued that one must aim to base one's reasonings on *verae causae*, and Darwin was desperately keen to show that his evolutionary reasonings were based on a *vera causa*, natural selection. But how was Darwin to show beyond doubt that natural selection was a *vera causa*? Here, Herschel's discussion becomes of vital interest: the most convincing evidence that something is a *vera causa*, Herschel argued, occurs when we can argue analogically from something else which we know already to be a *vera causa*. He wrote:

Here, then, we see the great importance of possessing a stock of analogous instances or phenomena which class themselves with that under consideration, the explanation of one among which may naturally be expected to lead to that of all the rest. If the analogy of two phenomena be very close and striking, while, at the same time, the cause of one is very obvious, it becomes scarcely possible to refuse to admit the action of an analogous cause in the other, though not so obvious in itself. For instance, when we see a stone whirled round in a sling, describing a circular orbit round the hand, keeping the string stretched, and flying away the moment it breaks, we never hesitate to regard it as retained in its orbit by the tension of the string, that is, by *a force* directed to the centre; for we feel that we do really exert such a force. We have here *the direct perception* of the cause. When, therefore, we see a great body like the moon circulating round the earth and not flying off, we cannot help believing it to be prevented from so doing, not indeed by a material tie, but by that which operates in the other case through the intermedium of the string, – *a force* directed constantly to the centre. (Herschel 1831, p. 149)

Take note, not only of the overall point Herschel was trying to make, but of his actual example. We have, in the case of the swinging stone, *a force, directly perceived and caused by us*; hence we know there must be a force causing the moon to swing around the earth. Darwin, who thought already of natural selection as a force, realized that he had an absolutely identical situation in biology. We have in artificial selection *a force directly perceived and caused by us*; hence, analogically, given the struggle and given wild variation, it cannot be denied that there is a natural force of selection making different organisms, just as man makes different organisms. In Herschel's own terms he had definitive proof that natural selection was a *vera causa*. No wonder therefore that Darwin, who had just finished reading Herschel, wrote excitedly that it was a virtue of his theory that his mechanism of evolutionary change was virtually the same mechanism as that of domestic organic change.

This, then was Darwin's major motive for stressing an analogy between artificial and natural selection. What we find, from his notebooks and his reading lists, is that in 1839 Darwin set with a will to the study of the results of breeders, to see just how much and how permanent artificial changes would be. To this end, Darwin was in a peculiarly favoured position. First, his family had long and with success kept and bred pigeons for a hobby (Meteyard 1871). So he had direct access to the world of the pigeon fancier. Second, his uncle (and after early 1839 father-in-law) Josiah Wedgwood (Jnr) had extensive experience in sheep breeding and selecting.[29] Wedgwood was much involved in the Society for the Diffusion of Useful Knowledge,[30] which was trying to disseminate breeding knowledge, in the main through the sponsorship of the classic works on breeding by Youatt, many of which works Darwin read in 1839. Convinced of the importance of the analogy between artificial and natural selections, Darwin had little trouble in persuading himself that, far from detracting from an evolutionary hypothesis, the domestic world supported it. The arguments of Whewell, and others like Lyell, were countered, and so, when Whewell in his 1839 presidential address to the Geological Society said 'If we cannot reason from the analogies of the existing to the events of the past world, we have no foundation for our science', Darwin replied triumphantly 'but experience has shown that we can and that analogy is sure guide and my theory explains why it is sure guide.'[31]

Until he felt ready to write out his theory, Darwin's chief concern was that of seeing how his mechanism of natural selection could be used to solve problems in different areas like instinct, geographical distribution, and so on. Although Darwin thought always that the analogy from artificial selection was support for his theory, the chief proof for him of its truth was that it had explanatory power in all of these many diverse areas.[32] And in this belief, as I have shown, Darwin was treading in the footsteps of Herschel and Whewell. There was, however, more to his acceptance of the Herschel-Whewell position on theory-confirmation than so far explained, and when this is revealed, an answer is found for one of the knottier puzzles in the Darwinian story, namely Darwin's attitude towards embryology.

In the *Origin* and in the *Sketch* and the *Essay*, Darwin used natural selection to explain one of the more curious aspects of embryology, namely that organisms of different species which have widely different adult forms sometimes have almost identical embryonic forms. Darwin pointed out that the embryos of different species are often under almost identical environmental conditions (and hence subject to the same selective forces), whereas the adults might be under very different conditions (and hence subject to very different selective forces). He suggested that some new variations may occur just in the adult forms, and that the different selective pressures on the adults could lead to different variations being preserved, and hence to the difference in adults of different species which we see today. The embryos of different species, on the other hand, would be subject to the same selective pressures, and hence there would be no parting of the ways. Obviously, absolutely crucial to this explanation is the belief that selection can preserve a characteristic for just *part* of an organism's life, in particular, in the adult form, with no selective pressure forcing it back to the embryonic form.

Darwin put inordinately great value on this embryological explanation. Although the discussion of embryology is one of the briefest in the *Origin*, time and again he argued that one must accept his theory, not merely because it explains so widely, but because it explains the facts of embryology. Thus, to Hooker he wrote that 'Embryology is my pet bit in my book, and, confound my friends, not one has noticed this to me' (Darwin 1887, 2, p. 244). And to Lyell he wrote that, if Lyell were not prepared to accept his theory, 'you give up the embryological argument (*the weightiest of all to me*), and the morphological or homological argument' (1887, 2, pp. 340-1, Darwin's italics). And although many people thought embryology important to evolutionary studies because, like Ernst Haeckel, they believed ontogeny in some sense 'recapitulates' phylogeny, for Darwin it was his own embryological explanation which was important:

> Hardly any point gave me so much satisfaction when I was at work on the *Origin* as the explanation of the wide difference in many classes between the embryo and the adult animal, and of the close resemblance of the embryos within the same class. (Darwin 1969, p. 125)

The reason why embryology was so important for Darwin stemmed directly from his acceptance of the Herschel-Whewell philosophy of theory-confirmation. We have seen that for Herschel (and for Whewell) the highest possible mark of theory-truth occurred when a theory explained phenomena, not built into the original explanation but perhaps even surprising or thought hostile by the scientist when he first conceived of his theory. 'Evidence of this kind is irresistible, and compels assent with a weight which scarcely any other possess' (Herschel 1831, p. 170). But in embryology Darwin had just such evidence. After he had thought of natural selection as an evolutionary mechanism, he was at first convinced that the struggle would be such that no characteristic could be preserved, unless it was of value for the whole life of an organism, 'No structure will last without it is adaptation to *whole* life of animal, and not if it be solely to womb as in monster, or solely to childhood, or solely to manhood – it will decrease and be driven outwards in the grand crush of population – ' (E, p. 9, his italics).[33] Then, some time after writing this (but before writing the *Sketch* of 1842), Darwin realized that his theory did not commit him to this, and that if he took the opposite position, he could explain the facts of embryology: facts which were certainly hostile to the theory if it did indeed imply that characteristics must be of value to the organism's whole life.

I suspect that Darwin's realization that he had in his grasp an explanation of the facts of embryology may have come through his study of domestic organisms. Breeders usually select just adults, with the consequence that although the juveniles of different varieties may be similar the adults are different. Darwin certainly used facts like these as strong evidence for his embryological explanation. But, be this as it may, embryology yielded just the kind of evidence Herschel and Whewell argued counted most decisively in favour of a theory's truth, and their follower Darwin agreed. The facts of embryology, which had once seemed hostile to his theory were now the theory's most positive evidence.[34]

Curiously, Darwin's feeling that he had definitive evidence in favour of his theory was not a feeling universally shared by the Darwinians. T.H. Huxley, Darwin's 'bulldog', always had reservations about the ultimate efficacy of natural selection, arguing that it would not be proven definitively as an evolutionary mechanism until someone had given direct evidence that selection, artificial or natural, had caused physiological speciation (Darwin 1887, 2, p. 198). In this he differed from Darwin who was absolutely convinced that because of his theory's wide scope (extending to embryology) it was conclusively proven. A plausible hypothesis of this difference of opinion lies in the fact that Huxley was much influenced by the philosophy of Mill (Ellegard 1958), and that the question of theory-confirmation is one point where we do find a difference between Herschel and Whewell on the one hand and Mill on the other. In particular, Mill (1843) denied that explanations of surprising or apparently hostile phenomena are incontrovertible marks of theory-truth (2, pp. 22-3). Although they may add to a theory's likelihood, argued Mill, they still leave room for doubt. Possibly, the difference between Darwin and Huxley reflects the difference between their philosophical mentors. Mill himself, although sympathetic to Darwin's theory, was certainly not convinced of its ultimate truth. (See Mill 1875, 2, p. 19.)

I come to the final aspect of Darwin's thought for which I want to suggest that the Herschel-Whewell influence was crucial. Darwin always thought that his theory of evolution was essentially incomplete. In particular, he thought he ought to provide a theory of heredity to explain the facts of new variation and transmission from one generation to the next. As is well known, eventually Darwin did produce such a theory, 'pangenesis', although he never incorporated it into the argument of the *Origin.* (See Darwin 1868.)

Why did Darwin feel the need to supply such a theory as pangenesis? Why did he not take the facts of variation as given? The reason lies in the Herschel-Whewell dichotomy between formal or phenomenal laws and causal or fundamental laws, and Darwin's acceptance of this dichotomy. Herschel and Whewell argued that one ought, as it were, get behind the phenomena to the underlying causes, where these might well involve invisible entities. Darwin agreed that one ought to get behind the visible facts of variation and delve into causes perhaps unseen. We find that in early 1839, as he was putting his theory together, he speculated on this problem *in Whewell's very language.*

> Those discovering the *formal* laws of the corelation of parts in individuals, will care little, whether the individual be species or variety, but to discover *physical* laws of such corelations, and changes of individual organs, must know whether the individuals forms are permanent, all steps in the series, their relation to the external world, and every possible contingent circumstance...
>
> Treating of the formal laws of corelation of parts and organs it may serve perfectly to specify types and limits of variation, and hence indicate gaps. – by this means the laws probably would be generalized, and afterwards by the examination of the special cases, under which the individual stages in the series have been fixed, to study the physical causes. (Darwin, E, pp. 53-5, his italics and spelling; written early November 1838.)

Darwin felt that one had to get to the physical laws of variation, and he did not rest until he produced his theory of pangenesis, a theory which traces the facts of heredity to minute *unseen* gemmules being cast off by body cells. Not surprisingly, although few of his friends felt much enthusiasm for this theory, Darwin defended it on the grounds that it explained a wide variety of phenomena (Darwin 1887, 3, p. 78). He remained faithful to the Herschel-Whewell philosophy.

EPILOGUE

Both Herschel and Whewell reacted unfavourably to the theory of the *Origin*. Herschel, as already noted, spoke of the law of higgledy-piggledy, and Whewell reputedly refused to let the offensive volume into the library of Trinity.[35] They both felt that Darwin had failed to do what any good biological theorist *must* do, pay adequate recognition to the role of God's Design in the formation of organisms. To them it was inconceivable that organic adaptation, something like the hand, had not been Designed, and to them it was inexcusable for a biologist not to have given this Design a central explicit place in his theorizing. Actually, the reactions of Herschel and Whewell did differ somewhat: Whewell rejected Darwin's theory entirely,[36] whereas Herschel thought Darwin's theory might be salvaged and shown of value, if only he would make place for Design. We must admit 'an intelligence, guided by a purpose.... On the other hand, we do not mean to deny that such intelligence may act according to law' (Herschel 1861, p. 12). Herschel went on to concede that, granting some reservations about the origin of man, 'we are far from disposed to repudiate the view taken of this mysterious subject in Mr. Darwin's book' (p. 12).

Darwin thought that his mechanism of natural selection made an appeal to explicit Design unnecessary. Organic adaptation could be seen to be the result of normal, undirected laws. Against Herschel, Darwin wrote 'astronomers do not state that God directs the course of each comet and planet. The view that each variation has been providentially arranged seems to me to make Natural Selection entirely superfluous, and indeed takes the whole case of the appearance of new species out of the range of science' (Darwin 1887, 2, p. 191).

Herschel and Whewell had presented to the world a philosophy of science inspired chiefly by Newtonian physics, particularly Newtonian astronomy. Their greatest pupil had learnt his lesson well; so well in fact that when the time came, Darwin, as pupils are wont to do, turned their teaching back against them. Darwin's was a theory modelled, through the medium of Herschel and Whewell, on Newtonian astronomy. Why then should his theory be expected to do that which astronomy does not do?

NOTES

1 The direct source of information about Darwin's life is his *Autobiography* (Darwin 1969).

2 Herschel's life and work is discussed briefly in Cannon (1961); Partridge (1966); Ducasse (1960a).

3 'Deduction' was a word used very loosely in the nineteenth century. One use was the modern use, the conclusion being contained in the premises, and it is clearly this sense that Herschel meant here. See also Whewell (1840), Aphorism XVI. 'In Deductive Reasoning, we cannot have any truth in the conclusion which is not virtually contained in the premises.'

4 Information on Whewell's life and work can be found in Todhunter (1876); modern discussions of Whewell's work include Ducasse (1960b); Butts (1968, 1970); Strong (1955); Buchdahl (1971); Laudan (1971).

5 By 'Newtonian astronomy,' here as elsewhere, I mean the astronomy of the 1830s.

6 See particularly book 7, chapter 4, 'Verification and Completion of the Newtonian Theory'. Section 3, 'Application of the Newtonian Theory to Secular Inequalities' deals explicitly with surprising or unanticipated phenomena.

7 Laudan (1971), suggests that there is a subtle difference between Herschel and Whewell on theory-confirmation. 'Whereas Whewell attaches greatest importance to the explanation of *surprising* facts, Herschel seems to lay greatest stress on the successful explanation of facts which had previously been regarded as counter-examples', (1971:374, n. 13).

8 Unpublished letter dated 5 February 1831, from Darwin to W. Darwin Fox, property of Christ's College, Cambridge.

9 Darwin filled four notebooks on the species problem between mid-1837 and mid-1839; that is, just at the time when he was discovering his theory of evolution through natural selection. These he labelled 'B', 'C', 'D', and 'E'. Darwin also wrote two (consecutive) notebooks which he labelled 'M' and 'N'. M was begun in mid-1838, and on the cover Darwin later wrote: 'This book full of Metaphysics on Morals and Speculations on Expression'. N was started on 2 October 1838 and was labelled 'Metaphysics and Expression'. I have worked out my dates from comments in these notebooks and from a reading list Darwin kept, 'Books to be Read'. All of these books are in the Darwin Collection, University Library, Cambridge, as are Darwin's copies

of Herschel's *Discourse* and *Astronomy* (1833, London: Longman) and Whewell's *History.* Comments about Herschel in the margins of the *History* suggest that Darwin read first Herschel's *Discourse* and then Whewell's *History.* There is a reference to the *History,* showing Darwin had finished reading it ('The end of each volume of Whewell's Induction History contains many most valuable references') on p. 69 of E, dated Dec. 16th (1838). On p. 49 of N there is a reference to Herschel's *Discourse,* written after Nov. 28th (the last preceding date, given on p. 45). Darwin read Herschel's *Astronomy* at some point – it is dated 1837. [Additional note added in 1988: The B, C, D, E notebooks were first transcribed and edited by G. de Beer (1960-7), and the M and N notebooks by P. Barrett in Gruber and Barrett (1974). Recently all have been re-transcribed and re-edited in one volume, Barrett *et al* (1987). Following convention, I refer here to Darwin's own pagination.]

10 Whewell wanted Herschel to take over presidency of the Geological Society.

11 Information from Henslow's lecture list, Darwin Collection.

12 My evidence for Whewell and Darwin's connections with the Geological Society comes from the contemporary minute books, still in the Society's possession. I am obliged to the Society for having been allowed to look at these.

13 Information from unpublished letters from Darwin to Whewell, Whewell Collection, Trinity College, Cambridge.

14 Unpublished letters in the Whewell Collection.

15 Letter from Darwin to Revd. J.S. Henslow, March 1834. In Barlow (1967), p. 87.

16 Darwin, 'Books to be Read', p.7; notebook C, p.72.

17 'A short time since I finished, having only skimmed parts before ... the History of Inductive Sciences' – unpublished letter from Darwin to Whewell, postmark April 17, 1839, property of Trinity. In notebook N, p. 14, dated Oct. 8th, there is a reference to the *History* 'V. Whewell, Induct. Science, vol. 1, p. 334'.

18 Darwin praised it to R. Brown (1969:104), and to Whewell in the unpublished letter of April 1839.

19 'Books to be Read', no page number given. See also Herschel (1841).

20 In the late 1830s, apart from his evolutionary theory, Darwin produced a coral-reef theory and an explanation of the parallel roads of Glen Roy. M.T. Ghiselin (1969) argues that Darwin was hypothetico-deductive in his coral-reef theory, and M.J.S. Rudwick informs me that Darwin aimed at a consilience of inductions in his Glen Roy discussion. [See also Rudwick 1974.]

21 See Whewell (1837) 3:620, and Lyell (1881) 2:5-6. In August 1838 Darwin did read with avid interest a review by Sir David Brewster of Comte's *Cours de Philosophie Positive.* What he would have got from this is that the aim of science is the positive stage

(Brewster 1838:280), that 'the fundamental character of *Positive Philosophy* is to regard all phenomena as subjected to invariable natural laws' (p. 281), and that the best of all laws is the Newtonian law of gravitational attraction (p. 282).

22 They both wanted organic origins through law, but both thought that these laws cannot be evolutionary. Whewell plumped for miracles.

23 In particular Wilkinson (1820) and Sebright (1809). They are referred to in notebook C:133. I discuss the pamphlets and their implications in Ruse (1975d).

24 Lyell, *Principles* 5th ed., 2: 430. (Darwin owned and read this edition.)

25 Wallace (1855) proposed the law that 'every species has come into existence coincident both in time and space with a pre-existing closely allied species'. He wrote 'Granted [this] law, and many of the most important facts in Nature could not have been otherwise, but are almost as necessary deductions from it, as are the elliptic orbits of the planets from the law of gravitation.'

26 The passage comes from the *Bridgewater Treatise* (Whewell 1833: 365).

27 The Wilkinson pamphlet, read mid-1838, suggests that some artificial change may be permanent, and Darwin noted this point. But even then, he still wrote (later) that the possible change in pigeons is limited.

28 This comment was made between Dec. 16th and Dec. 21st, 1838.

29 See correspondence with Thomas Poole (in British Museum). Wedgwood at one point had 2000 sheep, including 300 merinos, which he was trying to introduce to England.

30 He was president of the Etruria branch.

31 Darwin, notebook E, 128. Interestingly, Darwin sought information on breeding by means of a printed questionnaire – a technique explicitly advocated by Herschel (1831), p. 134.

32 Letter to G. Bentham, May 1863, (Darwin 1887, 3:25). In this letter, Darwin wrote of natural selection as a *vera causa*, although by then the term was used almost as loosely as 'deduction'.

33 This passage was written in October 1838.

34 Even if Laudan is right in finding a difference between Herschel and Whewell on theory-confirmation, Darwin's embryological explanation seems to fit either's criteria.

35 This is a story of Huxley's, and should perhaps be taken with a pinch of salt – although, knowing Whewell, not too big a pinch.

36 See Todhunter (1876), 2: 433-4; and Whewell's preface to the seventh edition of his *Bridgewater Treatise* (1863).

Chapter Two

CHARLES DARWIN AND GROUP SELECTION

INTRODUCTION

In recent years evolutionary biologists have shown much interest in the question of the levels at which natural selection can be said to operate (Lewontin 1970). Generally speaking, confining ourselves initially to the non-human world, it is probably true to say that although V.C. Wynne-Edwards in his *Natural Regulation of Animal Numbers* (1962) argued strongly for the wide-spread efficacy of some form of group selection, most evolutionists would agree with G.C. Williams' reply, *Adaptation and Natural Selection* (1966), in which it was argued that essentially selection must start with the individual. Nevertheless, a number of studies have been aimed at showing how under certain circumstances selection could work at the group level. Hence, it is probably true to say that matters are not yet definitively settled, either theoretically or empirically. (See, for instance, Levins 1970, Boorman and Levitt 1972, 1973, and Wade 1978.)

The debate has been given added zest by the fact that the assumption that selection almost invariably centres on the individual is crucial to the theories and conclusions of the sociobiologists (those biologists interested in animal social behaviour). (See Wilson 1975a, Trivers 1971, 1972, 1974, Barash 1977, Alexander 1971, 1974, 1975, 1977a, 1977b, Dawkins 1976, Hull 1978a, Ruse 1977c, 1978, 1979b.)

Indeed, what the sociobiologists claim, as a major feature of their work distinguishing it from that of earlier students of the biology of animal behaviour, is that they alone make the right choice of individual over group selection. This in itself would hardly be a matter of great controversy; but since most of the sociobiologists want to apply their theorizings from the animal world directly to the human world, inevitably there has been some rather heated discussion about whether one can properly use the notion of individual selection to explain the evolution and maintenance of all significant human behaviour. The critics of sociobiology feel that such an attempt leads to a reactionary distortion of human sociality, and they argue that other causes of human behaviour must be sought in explanation. (See Allen *et al.* 1975, 1976, 1977, Sahlins 1976.)

Of course, one does not have to be a supporter of some form of group selection in the non-human world to be a critic of human sociobiology. Nevertheless, some eminent biologists do fall into both categories, and moreover, I suspect they see important ideological links in their over critique of the all-sufficiency of individual selection. For instance, both Levins (1970) and Lewontin (1970) have allowed the possibility of group selection in certain special situations, they are both against sociobiology, and by their own admission they see the totality of their work as part of an overall Marxist-orientated biology (Lewontin and Levins 1976).

Because both the sociobiologists and their critics resort to the familiar tactic of trying to legitimize the present by reference to the past (one finds protagonists on both sides claiming that they alone stand in the true evolutionary tradition (Barash 1977, Sahlins 1976)) there is therefore some interest in seeing whether the debate about the levels of selection stretches back to the first announcement of the theory of evolution through natural selection, and where precisely the theory's chief formulator, Charles Darwin, stood on the matter. Such an historical enquiry is the aim of this paper. First, I shall consider Darwin's position on the levels of selection in his major work, *On the Origin of Species* (1859). Second, I shall see how his ideas developed over the next twelve years. Third, I shall look at his ideas in his major significant work on human beings, *The Descent of Man* (1871). Because there has been some confusion about what is meant by 'individual selection' (West Eberhard 1975, Maynard Smith 1976), I shall for the sake of clarity mean by 'selection' that which in some sense concerns an individual's reproductive interests. This could be directly through the individual, or indirectly in some way: for instance, by kin selection, where an individual's interests are furthered through close relatives; through parental manipulation, where a parent directs an offspring to its own interests; or through reciprocal altruism, where an individual is selected to do favours for others in the hope of returns (Ruse 1979b). By 'group selection' I shall mean selection in some way causing characteristics which help others, including non-relatives, in an individual's group, most probably by the species. There is not necessarily any hope of return for the individual.

ON THE ORIGIN OF SPECIES

The early chapters of the *Origin*, those in which Darwin introduced his mechanisms for evolution, certainly give the impression of a Darwin who was going to be firmly committed to individual selection. As is well known, he did not just present natural selection without any theoretical backing, as an axiom as it were. Rather, he argued first to a universal struggle for existence, from premises modelled on Malthus's ideas about available food and space. Then from the struggle, Darwin went on to argue that, assuming that there was heritable variation of the required amount and kind, the survivors and reproducers in the struggle would on average be different from the losers: natural selection (Flew 1959, Ruse 1971, 1975c). Now, as he introduced the struggle, Darwin gave strong evidence that he was going to be thinking at the level of the individual. To have some sort of group selection, one has got to minimize the tensions or rivalries within the group. When, for instance, the ethologist Konrad Lorenz (1966) invoked group selection, he did so because he was trying to show how it is that animals have mechanisms inhibiting all-out attacks on conspecifics. Darwin, however, saw the struggle (which was going to lead to selection) as acting just as much between conspecifics as between any two organisms:

> As more individuals are produced than can possibly survive, there must in every case be a struggle for existence, either one individual with another of the same species, or with the individuals of distinct species, or with the physical conditions of life. (Darwin 1859, p. 63)

In fact, he then went on to say that the closer the relationship the more severe the struggle: 'the struggle almost invariably will be most severe between the individuals of the same species' (p. 75). It should be added, of course, that he was not necessarily thinking of struggle in the sense of hand-to-hand combat, but struggle for resources of various kinds and so forth. Nevertheless, the point remains that he was viewing the crucial biological tensions as much within the group as without.

Coming to selection itself, we find the same emphasis on the individual. For instance, to illustrate how natural selection might work Darwin gave the imaginary example of a group of wolves, hard-pressed for food (1859, p. 90). He suggested that the swiftest and slimmest will be selected, because it will be they alone who will catch the prey. Hence, there will be evolution towards and maintenance of fast, lean wolves. Obviously, the crux of this explanation is that some wolves survive and reproduce whereas others do not. There is no question here of selection working for a group; rather it is all a matter of individual against individual.

Although an example like this shows that Darwin thought natural selection itself to be individual orientated, his commitment to the individual is perhaps best illustrated by his variant form of selection, sexual selection (Darwin 1859, p. 87; see also Ghiselin 1969, Mayr 1972). This Darwin divided into two forms: through male combat, where the males compete between themselves for the females, and through female choice, where females choose between males displaying in various ways. It is true that Darwin was criticized for this mechanism, particularly on the grounds that female choice anthropomorphically supposes that animals have the same standards of beauty as humans.[1] But this is as it may be. What is important is how clearly sexual selection shows that Darwin was thinking of selection as something that could act between fellow species members, preserving a characteristic that gives an organism an advantage over conspecifics. There is no place here for the preservation of characteristics of value to conspecifics at the expense of the individuals within a group.

Introducing selection and its foundations, therefore, Darwin gave the impression that he was going to be a fairly rigorous individual selectionist. But was he completely committed to individual selection? Did he feel that there could ever be a case where selection could and would act for the benefit of the group? In the *Origin*, as we turn from Darwin's introduction of his mechanisms to their applications, we find that with respect to the levels of selection dilemma, there are two points at which Darwin had to make a decision: when he discussed social insects and when he discussed hybrid sterility. Let us take them in turn.

By 'social insects' is meant insects with sterile castes, all living together in a community. The problem is how one explains the sterility of individuals in some of these castes, and how members of sterile castes could have evolved to be so very different from fertile fellow community members. Surely one must invoke some sort of group selection to explain how the sterile community members, so helpful to the group, so unhelpful to themselves, evolved? About the question of sterility, Darwin wrote as follows:

> How the workers have been rendered sterile is a difficulty:
> but not much greater than that of any other striking
> modification of structure; for it can be shown that some
> insects and other articulate animals in a state of nature
> occasionally become sterile; and if such insects had been
> social, and it had been profitable to the community that a
> number should have been annually born capable of work,
> but incapable of procreation, I can see no very great
> difficulty in this being effected by natural selection.
> (Darwin 1859, p. 236)

In a related fashion he tackled the problem of how the members of some sterile castes can be very different from their parents, and different from the members of other castes. Drawing on a favourite analogy of the effects of artificial selection in the domestic world, Darwin pointed out how breeders can work indirectly, raising desired organisms (which are killed without reproducing), because the fertile parents can somehow latently carry the characteristics of these organisms. Similarly, in the wild, fertile parents could have sterile offspring who help the community, and also could pass on to their fertile offspring the potential to have such sterile offspring in turn. And if different kinds of sterile offspring could be used in a community, these too could be formed by selection (Darwin 1859, pp. 237-8).

There is no group selection here, where 'group selection' is understood in our above-defined sense as involving unreturned aid to non-relatives. The key to Darwin's argument is that the sterile altruists are closely related to the fertile members of the community. The sterility occurs because it is of value to all the related members of the nest, and also, Darwin thought, in part because it prevents interbreeding and the production of less effective hybrid forms (that is, forms, less effective for the community). But selection is not preserving characteristics exclusively of value to non-relatives.

One should nevertheless add that although Darwin was certainly an individual selectionist at this point, even one sufficiently sophisticated to see how individual selection can work through a closely related community, he did not really bring his argument to the level brought by today's sociobiologists. Some sociobiologists today argue that caste differentiation (in the Hymenoptera at least) is a function of kin selection: because of the haploidiploid method of sex determination in the Hymenoptera, sisters are more closely related to sisters than to daughters, and thus there is a genetic advantage in raising sisters (Hamilton 1964a,b, Trivers and Hare 1976). Other of today's sociobiologists argue that the key is parental manipulation, where parents make some offspring sterile altruists towards siblings (Alexander 1974). Both of these explanations see that there can be reproductive conflicts between community members, despite the close relationships. Darwin however, saw the community members united in common interests. It would therefore be somewhat anachronistic to say that Darwin was a kin selectionist rather than a parental manipulator, or vice versa. Because of his ignorance about the proper principles of genetics, his analysis was not as precise as that. One might be tempted to say that Darwin was a shade closer to group selection than today's sociobiologists, because he saw no conflict between relatives, despite their lack of genetic identity. Perhaps so – but again I am inclined to think that one would be anachronistically reading into Darwin's work something more subtle than is really there.

The other place in the *Origin* where Darwin might have been tempted towards a group selection mechanism was over the question of sterility; either the sterility between members of different species, or if hybrids were formed, the sterility of these hybrids (Kottler 1976). A priori one might think that the very usual sterility between species or the sterility of hybrids (for instance, the sterile horse-donkey, the mule) would have been something fashioned by selection. If there are two forms, each adapted to its respective environment, a hybrid would be (literally!) neither fish nor fowl. Hence, it would be of advantage that such a hybrid either be barred altogether or if possible be sterile because it could not then reproduce and give rise to ➤ further ill-adapted forms. The problem is, of course, to whom the absence or sterility of the hybrid would be of advantage. If the hybrid is not formed at all, then the parents lose any chance of offspring. If the hybrid is formed, then it would clearly not be of direct advantage to the hybrid itself to be sterile. Nor, differentiating this case from that of the social insects, would it be of advantage to the parents that the hybrid offspring be sterile – there is no question of the hybrid being freed through its sterility to aid its related community. At best the advantage from absence or sterility of hybrids would be to the parent species, who would thus gain better evolutionary prospects because no energies or resources would be going into ill-adapted hybrid offspring.

But to Darwin, apart from the fact that he could not see why in nature one gets so many degrees and forms of sterility (assuming selection does cause sterility), the unambiguous group selection required to cause sterility was apparently just not a live possibility. Although he did not provide a detailed theoretical attack on group selection, Darwin clearly hinted that he could not see how group selection, favouring the group over the individual, could work at all. Further, he went to some pains to show how individuals of different species frequently cannot interbreed at all because of incidentally formed differences, and similarly how hybrid sterility is an incidental fact, brought about by the lack of harmony between the different contributions by the parents to the hybrid's reproductive mechanisms:

> On the theory of natural selection the case [of sterile hybrids] is especially important, inasmuch as the sterility of hybrids could not possibly be of any advantage to them, and therefore could not have been acquired by the continued preservation of successive profitable degrees of sterility. I hope, however, to be able to show that sterility is not a

specially acquired or endowed quality, but is incidental on other acquired differences. (Darwin 1859, p. 245)

BETWEEN THE *ORIGIN* AND THE *DESCENT*

There are two items of particular interest in the 1860s. First, there is the fact that Darwin himself wrestled at length with possible selective causes of sterility. Second, relatedly, Darwin and natural selection's co-discoverer, Alfred Russel Wallace, debated the individual-group selection problem. By the end of the decade, with respect to the animal and plant worlds, there was nothing implicit about Darwin's commitment to individual selection. He had looked long and hard at group selection and rejected it. Let us take in turn the matters which engaged Darwin in the levels of selection problem.

Almost immediately after the publication of the *Origin*, Darwin's interests turned to botany. Amongst other plants that he studied were members of the primula family: primroses and cowslips (Darwin 1862; see also Ghiselin 1969, Whitehouse, 1959). Members of this kind of species come in one of two different forms. Some have long styles, with the stamens tucked right away down the tube of the corolla; others have short styles, with the stigma right down the tube and the stamens at the mouth (see Figure 2.1). Hitherto, these different forms had been considered accidental varieties, but Darwin was able to show that the two forms play important roles in the cross fertilization of the (hermaphroditic) plants. In particular, the crosses between plants of different types are far more fertile than crosses between plants of the same type (see Figure 2.2).

In the different degrees of fertility between the different kinds of crosses, Darwin thought that he found a clue pointing to the possibility, in some special cases at least, that selection may have had a hand in sterility barriers. Why should a plant be barred from reproducing in a satisfactory way with half of its fellow species' members? Darwin hypothesized that this barrier may be a consequence of a selection acting to prevent something which is known to be positively deleterious, namely self-fertilization (which produces inferior, inbred offspring). Plants, Darwin supposed, get selected for self-sterility, and then somehow

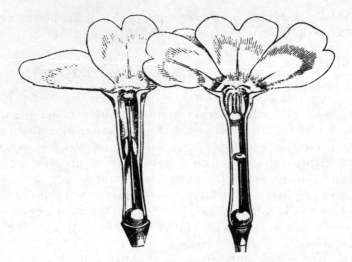

Figure 2.1 Long-styled (left) and short-styled (right)

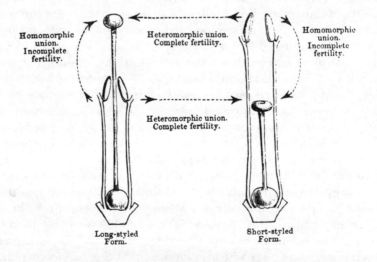

Figure 2.2 Heteromorphic and homomorphic unions
Source: C. Darwin 'On the two forms of dimorphic condition, in the species of *Primula,* and on their remarkable sexual relations', *Journal of the Proceedings of the Linnaean Society (botany)* (1962) 6, pp. 77-96

accidentally (at least, not through selection) this gets transferred into a sterility with all forms like the plant itself. Furthermore, thought Darwin, if this hypothesis be true, we should find that the kind of plant which stands the greatest chance of being self-fertilized has the greatest barriers to prevent it. Since the primulae are fertilized by insects, it is obviously the short-styled form which runs this greater risk, and Darwin was happy to be able to confirm that short-short crosses were indeed even less fertile than long-long crosses. He therefore concluded:

> Seeing that we thus have a groundwork of variability in sexual power, and seeing that sterility of a peculiar kind has been acquired by the species of *Primula* to favour intercrossing, those who believe in the slow modification of specific forms will naturally ask themselves whether sterility may not have been slowly acquired for a distinct object, namely, to prevent two forms, whilst being fitted for distinct lines of life, becoming blended by marriage, and thus less well adapted for their new habits of life. (Darwin 1862, p. 61)

Judging from this passage alone, one might think that Darwin had now turned, not merely to a selection hypothesis for sterility, but to a group selection hypothesis, at least in partial explanation of sterility in those organisms which are hermaphroditic. (This latter clause about hermaphrodites was not necessarily that great a restriction, because his work on barnacles had convinced him that sexual organisms evolved from hermaphrodites. See Darwin 1851, 1854, Ghiselin 1969, Ruse 1979a.) However, looking at his explanation carefully, it is clear that Darwin had turned to nothing of the sort. In as much as selection was supposed to cause sterility, it was for the good of the individual: so that it would not fertilize itself, thus causing only inferior inbred offspring. In so far as sterility was being generalized from the individual to the group, it was accidental, in the sense of not being of selective value. There was no question of selection for the group, however much Darwin's rather sloppy use of language might hint otherwise.

At the end of the above-quoted passage, where Darwin had speculated that selection might cause sterility, wisely he added reservations: 'But many great difficulties would remain, even if this view could be maintained' (Darwin 1862, p. 62). It was perhaps just as well that he covered himself here, for it was not very long afterwards that he discovered evidence which destroyed his hypothesis about the selective origin of sterility (Darwin 1865). The plant *Lythrum salicaria* has three forms, and they are all involved in cross-fertilization of members of the species. In particular, any one of the three forms (long-, mid- and short-styled) needs one of the other forms to effect the most efficient fertilization. If Darwin's hypothesis was correct, then since the closer the stigma and the stamens the greater the chance of self-fertilization, the closer the stigma and the stamens on two plants being crossed the greater should be the sterility barrier (excluding of course, cases where stigmas and stamens were in exactly the same positions, because this could not happen on a single plant). However, in fact it turned out that sterility was a direct function of the distance between stigmas and stamens; the very opposite to what was predicted in the hypothesis. Hence, Darwin was led to reject his short-lived speculations, and to return to his original position: sterility was an accidental by-product of individual selection.

Perhaps as a warning to others, in the fourth edition of the *Origin* (published in 1866), Darwin inserted a somewhat stronger discussion than hitherto as to why selection could not cause sterility, although like most of us, he found it difficult to reject a good idea – even if it was fairly clearly false; witness his reluctance to throw out his marine explanation of the parallel roads of Glen Roy (Rudwick 1974). Hence, in this edition of the *Origin* he did present his hypothesis about sterility in the primulae, but then he concluded that selection could not cause sterility (Darwin 1959, p. 446)!

In support of his strengthened claim that selection could not cause sterility, Darwin gave three reasons. First, he pointed out that selection could not be a necessary condition for sterility, because there are indisputable cases of species having developed in different geographical regions, and yet, although selection could not possibly have made them inter-sterile, they prove to be so when they are brought together. Second, there are cases where only one of the possible crosses between members of species proves sterile, and where the other cross is quite fertile (that is, male of species A crossed with female of species B is sterile, but male of species B crossed with female of species A is fertile).

Darwin could not see how selection could cause this asymmetrical relationship. Third and most important, he stated categorically that even if sterility is of value to the group, it is not of value to the individual, and it is at the level of the individual that selection operates. Sterility could not have been caused by natural selection, 'for it could not have been of any direct advantage to an individual animal to breed poorly with another individual of a different variety, and thus to leave few offspring; consequently such individuals could not have been preserved or selected, (Darwin 1959, p. 444). And in the 5th edition of the *Origin*, he added that if hybrids are born and they are less than fully fertile, then selection will act against them too. In short, he showed that he had thought consciously about and rejected group selection.

If indeed, with his full discussion of the reasons why sterility could not be fashioned by selection, it was Darwin's aim to warn others against thinking that selection could fashion anything (including sterility) of value to the group rather than the individual, his actions had exactly the opposite effect to that which he intended! In particular, his discussion (which he repeated practically verbatim two years later in his *Variation of Animals and Plants under Domestication* (1868)) spurred Wallace to react by embracing whole-heartedly the cause and efficacy of the group selection:

> It appears to me that given a differentiation of a species
> into two forms, each of which was adapted to a special
> sphere of existence, every slight degree of sterility would
> be a positive advantage, not to the individuals who were
> sterile, but to each form. If you work it out, and suppose
> the two incipient species *a...b* to be divided into two
> groups, one of which contains those which are fertile when
> the two are crossed, the other being slightly sterile, you
> will find that the latter will certainly supplant the former
> in the struggle for existence ... (Letter from Wallace to
> Darwin, 1868, in Darwin and Seward 1903, 1: 288.)

Although Darwin and Wallace each tried as politely as possible to see the viewpoint of the other, the beginning was basically the end of the matter also. Darwin reiterated that he could not see how sterility, so disadvantageous to the individual, could be preserved by selection; Wallace reiterated that he could not see

45

how sterility, so advantageous to the group, could fail to be preserved by selection; and that was that (Darwin and Seward 1903, 1, pp. 288-97). It is true that Darwin and Wallace were not entirely opposed; they both agreed that a disinclination to cross with members of another species could have been acquired by selection, even though Wallace, unlike Darwin, wanted to link this disinclination to sterility, should a cross ever so occur. But of course for Darwin this admission did not demand compromise on the matter of individual selection. Given a choice between hybrid offspring and offspring entirely of one's own species, it would certainly be to an organism's reproductive advantage not to waste effort on producing hybrids, unadapted to either parent's ecological niche. Hence selection could help one avoid producing such hybrids in the first place. The point was, as Darwin saw it, if the coupling had in fact taken place, it would not then be in an individual's advantage to promote sterility. In other words, if an individual had mated with another, it would not then be in its interests to yield less than fully fertile offspring.

And so we find Darwin and Wallace divided. Rather sadly Darwin concluded 'We shall, I greatly fear, never agree' (Darwin and Seward 1903, 1, p. 296). Wallace gallantly conceded that it was probably he himself who was wrong (although this did not stop him later in the century from repeating his own position in print! (Wallace 1889)). Wallace feared only that the problem of sterility 'will become a formidable weapon in the hands of the enemies of Natural Selection' (Darwin and Seward 1903, 1, p. 297). Incidentally, it is interesting to note that in this disagreement there are faint echoes of the other matter which separated Darwin and Wallace at this time: sexual selection through female choice (Marchant 1916, pp. 151-4). Darwin wanted to argue that the beauty of, say, the peacock as opposed to the peahen, is a function of the females choosing beautiful males. Wallace argued that the difference is essentially a function of the females being more drab than the males, this drabness coming through the female's need for camouflage from predators as they incubate their eggs and care for their young. In arguing this way, Wallace was certainly not invoking group selection. However, unlike Darwin, who was emphasizing the individual nature of selection by seeing the main competition (at this point) as coming from within the species, Wallace was de-emphasizing competition within the group by seeing the threat coming from without.

Concluding this section dealing with the years immediately following the *Origin*, we see therefore that Darwin continued to think about the problem of the proper level of selection, and that he became even more convinced that in the non-human world selection acts at, and only at, the level of the individual. Let us see now what happened when Darwin turned his attention to human beings.

THE DESCENT OF MAN

As is well known, what made Darwin's speculations in the *Origin* so unpalatable to so many were the obvious implications for man. As the geologist Charles Lyell sadly wrote to Darwin:

> It is small comfort or consolation to me, who feels that Lamarck or Darwin have lessened the dignity of their ancestry, making them out to be with.[t] souls, to be told, 'Never mind, you will be succeeded in unbroken lineal descent by angels who, like Superior Beings spoken of by Pope, "Will show a Newton as we show an ape".' (Wilson 1970, p. 382)

In fact in the *Origin* itself, Darwin hardly mentioned man. Wisely deciding not to draw more controversy than he need, Darwin deliberately restricted himself to a single final comment: 'Light will be thrown on the origin of man and his history' (Darwin 1859, p. 488). And this, surely the most understated claim of the nineteenth century, Darwin added only so that he would not later be accused of dishonourably concealing his own true beliefs. But his reticence should not be confused with indecision. It is clear, from private notebooks that he kept, that right from the time when he first became an evolutionist in the late 1830s, Darwin considered man as an animal on a par with other animals (Barrett *et al*, 1987). Indeed, the first hint that we have of him using natural selection as an evolutionary mechanism is a speculation about how selection might have improved human intelligence (Gruber and Barrett 1974, notebook M, p. 42).

Through the years, nothing at all changed, and so we find that when Darwin did finally write fully and publicly on man, in 1871 in his *Descent of Man*, he tried as carefully and thoroughly as he could to show that man evolved from other animals, by the same processes as held throughout the organic world. For instance, 'in the rudest state of society, the individuals who were the most sagacious, who invented and used the best weapons or traps, and were best able to defend themselves, would rear the greatest number of offspring' (Darwin 1871, 1, p. 196). The one thing which is perhaps noteworthy about Darwin's treatment of man, taken generally, is the very significant role which he gave to the action of sexual selection. And even this, to a certain extent, was forced on him by external circumstances, namely the apostasy of Wallace.

In his early years, Wallace had had even less religious belief than Darwin, and in fact when Wallace first started to write on man, after the *Origin* was published, he still treated him as a natural object. However, towards the end of the 1860s, chiefly because of a growing involvement with spiritualism, Wallace came to believe that there were aspects of human development which call for a creative power above any natural process of selection (Wallace, 1869, 1870; see also Kottler 1974, Smith 1972). For instance, he argued that only through some kind of supernatural interference could one explain the relative hairlessness of members of the human species. To counter Wallace, although undoubtedly also as a natural development of ideas which he had had previously, Darwin included in the *Descent* a very large general discussion of sexual selection, and then he argued that many of the differences between humans, both between the sexes and between different races, are due to this kind of selection: men struggle for the women they want, women are attracted to the dominant men, and so forth. Thus, something like human hairlessness can be explained as a function of early men finding hairy mates distasteful (Darwin 1871, 2, pp. 376-7).

The precise details of Darwin's general explanation of man's evolution are not our concern here. What is important to us is the obvious fact that normally he saw the individual man or woman as being the crucial unit in the selective process. There was no question that, when faced with his own species, he was going to swing round suddenly and start to argue as a general policy that for *Homo sapiens* alone the group, particularly the species, is the key element in the evolutionary mechanism. As we saw in the above quotation about the evolution of intelligence, it is when some individual man is brighter than his fellows that we get the important evolutionary consequences, for then it is that he (not everyone) will have an increased crop of children. Furthermore, whether Darwin was indeed right in giving sexual selection so important a role in human development, the fact is that he did emphasize in even greater detail the extent to which he saw evolutionary competition occurring within the human species. As pointed out earlier, by its very definition sexual selection takes place within a species, pitting conspecific against conspecific.

Nevertheless, in dealing with man's evolution there was one point (a point incidentally noted by Wallace as inexplicable through selection) where Darwin for once did quaver in his commitment to individual selection. This was over the evolution of the human moral sense: human awareness of and actions upon what is right and wrong. Darwin was certainly not about to follow Wallace in concluding that human morality implies that there must be a supernatural power guiding human morality; but for once he did lose sight of the individual and allow that possibly the unit of selection may have been the group, specifically the tribe. Rhetorically, Darwin asked: 'how within the limits of the same tribe did a large number of members first become endowed with [their] social and moral qualities, and how was the standard of excellence raised?' (1871, 1, p. 163). And then immediately he expressed his worries about the power of individual selection to bring about such morality:

> It is extremely doubtful whether the offspring of the more
> sympathetic and benevolent parents, or of those which
> were the most faithful to their comrades, would be reared
> in greater number than the children of selfish and
> treacherous parents of the same tribe. He who was ready
> to sacrifice his life, as many a savage has been, rather than
> betray his comrades, would often leave no offspring to
> inherit his noble nature. The bravest men, who were always
> willing to come to the front in war, and who freely risked
> their lives for others, would on an average perish in larger
> number than other men. (Darwin 1871, 1, p. 163)

In short, it would seem that natural selection working at the level of the individual could not bring about or preserve a heritable moral sense.

Of course, asking rhetorical questions and setting forth the difficulties do not in themselves imply absolutely that Darwin believed that the only way in which one could explain the human moral sense was through some sort of group selection; that selection preserved a feeling for morality because the moral group was more fit than the immoral group, even though the moral individual may have been less fit than the immoral individual. However, shortly after the just-quoted passages, Darwin did give evidence that this was the way in which he inclined:

> It must not be forgotten that although a high standard of morality gives but a slight or no advantage to each individual man and his children over the other men of the same tribe, yet that an advancement in the standard of morality and an increase in the number of well-endowed men will certainly give an immense advantage to one tribe over another. (Darwin 1871, 1, p. 166)

Moreover, apparently Darwin told one of his young followers that in the case of man's moral sense, he believed that selection had to be acting at the level of the type rather than the individual (Romanes 1895, p. 173).[2]

It would seem therefore that although Darwin resolutely opposed group selection in the non-human world, when it came to our own species, although again for almost everything he was an individual selectionist, in one crucial respect of our culture – our morality – he weakened and allowed that selection must have acted at the level of the population. He could not see how helping our fellows simultaneously helps our own reproduction, unless we make the reference unit of selection the group rather than the individual. Apparently, at the final point of evolution, Darwin became a group selectionist.

Nevertheless, whilst one can hardly deny some truth to this conclusion, there are two modifying points which should be made. First, it must be noted that even if Darwin became something of a group selectionist, he was never a group selectionist thinking that the crucial unit of selection is the species. His concern was at most for the tribe, and he was quite explicit that morality as it developed was to benefit fellow tribesmen, to the detriment of other members of the human species:

> There can be no doubt that a tribe including many members who, from possessing in a high degree the spirit of patriotism, fidelity, obedience, courage, and sympathy, were always ready to give aid to each other and to sacrifice themselves for the common good, would be victorious over most other tribes; and this would be natural selection.
> (Darwin 1871, 1, p. 166)

It is true indeed that Darwin allowed that as civilization rises, one's moral concerns extend, through the human world and even to animals (Darwin 1871, 1, p. 103). But it is probable that at this point he would have thought we would have left the strictly biological, and have entered the realm of what we might anachronistically call 'cultural evolution'. In other words, here Darwin would possibly have thought that the biological individual-group selection debate was irrelevant. Certainly, he thought that modern culture transcends the biological; for instance, he rather lamented the fact that modern medicine allows the infirm to survive and reproduce, because as he pointed out, biologically this leads to the race becoming less fit (1871, 1, p. 168).

It should also be noted that Darwin saw many of the tribe-members as being related (Darwin 1871, 1, p. 101), and he was quite clear that, as in the case of social instincts, human virtues can be spread through relatives, even if an individual does not survive. I am not suggesting that Darwin went so far as to think that the only kind of group selection involved in promoting morality collapsed into an individualistic kin selection. It is fairly definite that his primitive morality was supposed to aid all fellow tribesmen, including non relatives. But, it does seem fair to say that his group selection was of a rather mild variety.

The second modifying point about Darwin's acceptance of group selection is that, whatever its degree, it was hesitant at best. Indeed, along with his group-selection explanation of human morality, he offered an individual-selection explanation as well! He argued that morality could have come about through what today's sociobiologists would call 'reciprocal altruism', namely a form of enlightened self-interest, where being nice to others pays from an evolutionary viewpoint because they in turn are nice to you (Trivers 1971). Morality, Darwin suggested, may have begun because 'as the reasoning powers and foresight of the members became improved, each man would soon learn from experience that if he aided his fellow-men, he would commonly receive aid in return' (1871, 1, p. 163). In other words, it would seem that even as Darwin strayed from the pastures of individual selection to those of group selection, he checked himself. Hence, with respect to human morality he ended up sitting firmly on the fence!

CONCLUSION

Let us conclude our discussion by linking things to the contemporary scene. This is worth doing, not to make Darwin seem more modern than he really is, a pseudomember of the Harvard biology department – he is too great to need this revisionist treatment – but because as we saw earlier, participants on both sides of today's debate about the levels of selection, particularly as they pertain to the evolution of humans, have invoked the past in defence of their own positions and criticism of their opponents.

In the light of our discussion, one can only suspect that Darwin's sympathies today would lie with those who push individual selection a very long way. In the non-human world Darwin was a firm, even aggressive, individual selectionist. This did not of course stop him from arguing that peculiar phenomena like the social instincts would have evolved through such selection. However, as pointed out, his views on heredity were not sufficiently sophisticated for us to guess which of the various modern hypotheses about the evolution of insect sociality he would have favoured. We cannot, for instance, tell whether Darwin would have supported kin selection or parental manipulation as the true cause of hymenopteran sociality. But queries at this level do not negate the fact that for organisms other than man, he unequivocally invoked individual selection.

Given the facts covered in the last section, it is obvious that at least a slight gap starts to open up between the Darwin of the last century and the total believer in individual selection of this century. For almost all aspects of man, indeed, there would be agreement in principle about the power of the individual selection, even though there might possibly be differences about the specific workings of such selection. However, it must be admitted that with respect to the evolution of morality Darwin seems to have been more sympathetic to group selection than would be a modern extremist like (for instance) R.L. Trivers. Nevertheless, given Darwin's general commitment to individual selection, his acceptance of group selection for morality seems to have been motivated more by the negative cause of being unable exactly to see how individual selection can cause morality, than by the positive cause of thinking that group selection validates itself on its own merits. Thus, were Darwin to have seen modern work like Trivers's explanation of morality through individualistic reciprocal altruism, he might well have responded positively; particularly since he himself gave the rudiments of such a reciprocal altruism argument!

It might be added moreover that, if one is determined to see Darwin in a modern light, even some of the most notorious human sociobiologists seem to allow that not everything is completely typical when we come to morality. Wilson, for instance, concedes that for human culture the genes have 'given away most of their sovereignty' (1975a, p. 550), and he certainly thinks we can (and should) act morally in the interests of our group, rather than the individual. Otherwise he could not argue as he does about the need to eliminate the human population explosion (Wilson 1975b). Similarly, R.D. Alexander (1971) allows that we are now at the point where individual and group interest coincide, with respect to certain moral questions. In other words, the individual-group selection tension vanishes, because the two modes of selection fuse together. We know that Darwin thought there comes a time when humans (in some respects) escape their biology. Hence, he would undoubtedly feel fairly sympathetic to the general kind of position that these biologists try to sketch out.

Finally, let us offer solace to the opponents of human sociobiology. If one is uncomfortable with a rather extreme individual selectionism, particularly as applied to man, and if one yet wants historical precedent to legitimize one's yearnings, then no less than the sociobiologists can one find the most respectable of intellectual ancestors. One may not be able to claim one of the fathers of evolutionism, but one can claim the other: Alfred Russel Wallace. He was a group selectionist, and moreover he was not prepared to see man treated on a par with other organisms. I certainly do not want to pretend that today's biologists would find convincing the details of Wallace's doubts about the all-sufficiency of individual selection, or that those who criticize human sociobiology grind the same metaphysical axe as did Wallace (although interestingly, politically Wallace (1900) was fairly left-wing, as are many of today's critics). But, given Wallace's conclusions, it does seem true to say that the critics of human sociobiology are no less part of the evolutionary tradition than those they criticize!

NOTES

1 Such criticisms are still being made. See G. Himmelfarb (1962).
2 It must be noted, however, that Romanes had his own special views on speciation, which might have made him a less-than-reliable reporter: see Kottler (1976).

Chapter Three

WHAT KIND OF REVOLUTION OCCURRED IN GEOLOGY?

The one thing upon which we can all agree is that just a short while ago a major revolution occurred in the science of geology. Geologists switched from accepting a static earth-picture to endorsing a vision of an earth with its surface constantly in motion. (See Cox 1973, Hallam 1973, Marvin 1973, Wilson 1970.) It is true that early in this century the German geologist Alfred Wegener (1915) argued that the continents as we today find them have 'drifted' to their positions from other positions widely different. However, other than amongst a number of scientists drawn almost exclusively from the southern hemisphere, his ideas fell on deaf – or more precisely, contemptuous – ears. Then in the mid 1960s, almost literally overnight, the geological community swung around and embraced the hypothesis of continental drift, or what we shall see is perhaps more accurately called 'plate tectonics'.

Given the fact that the major topic of debate amongst philosophers of science in the past couple of decades has been the exact nature of a scientific 'revolution', one might think that so dramatic a revolution so close at hand, in a science which is really not particularly technical (at least is not as incomprehensible to the outsider as modern particle physics), would have attracted immediate and detailed attention by the philosophical fraternity. This however would have been to reckon without the average philosopher's obsession with white swans, black ravens, and red herrings, and his distaste for anything vaguely resembling real science. (By saying 'his' here, I am not being quite as sexist as I sound; female philosophers of science have shown a refreshing interest in what real scientists really do.) The revolution in geology has been greeted by philosophers of science with an absolutely crushing silence. To the best of my knowledge, not one of the major, or minor, journals in the field has mentioned continental drift and plate tectonics, let alone discussed them. (In contrast, there are at least two histories of the geological revolution, not to mention collections of the seminal contributions: Cox 1973, Hallam 1973, Marvin 1973.)

As might be expected, the real losers from this indifference are we philosophers, not the geologists. The geological revolution is exciting and dramatic, and it holds rich rewards for those of us concerned to understand the temporal development of science and the reasons making scientists change their minds. Rather than argue this case in the abstract, I shall show the validity of my argument and the truth of my conclusion by attempting what must necessarily be a preliminary and sketchy philosophical analysis of the geological revolution. Like almost every other philosopher of science, until very recently I had only the haziest notion of what had gone on in modern geology; hence, what I have to say will, I am sure, be riddled with errors. But if I can at least kindle the interest of others I shall feel satisfied, for I am sure that as they read into the literature they will grow to agree with me that modern geology is something which ought to figure largely in philosophical discussion.

On a more personal level I must add that I have found my excursion into the philosophy of geology particularly satisfying. For a number of years now I have been studying British philosophy of science of the nineteenth century, particularly its fine flowering of the fourth decade as manifested by the work of John F. W. Herschel and William Whewell. As might be expected from a famous astronomer and a noted tidologist, for these philosophers the paradigm science was Newtonian astronomy. However, as might also be expected from a time when geology was the fashionable science and from a decade whose opening was marked by the publication of Charles Lyell's magisterial *Principles of Geology* (1830-3), the philosophers of science showed a keen interest in the methodological and foundational problems arising out of geology (see Ruse 1976a, 1976b, 1979a). To me therefore, the opportunity to direct philosophy of science even a little back to its earlier areas of interest becomes peculiarly satisfying.

THE GEOLOGICAL REVOLUTION

The west coast of South America and the east coast of Africa surely do look awfully similar, and so it is not very surprising that almost as soon as the major features of our globe were mapped out there first occurred hints and suggestions that the continents have not always been fixed in their present positions, but have moved or 'drifted' from positions quite different: S. America and Africa originally being part of some primeval continent and having cracked apart. This hypothesis certainly never found universal favour, although most geologists seem to have agreed that things today are not as they used to be in the past. Apart from anything else, from Lyell onwards there were all sorts of suppositions of now-vanished land bridges between continents to explain present organic geographical distributions.

It was not until the beginning of this century that a really impressive case was made for continental drift. This came from the pen of the German scientist, Alfred Wegener. Arguing from a broad spectrum of evidence – the fit of the continents, organic distributions, the fact that the earth's surface seems to have two average levels pointing to the permanence of the continents and the sea (see Figure 3.1) – Wegener argued that today's continents evolved by drifting from some original super-continent, 'Pangaea' (see Figure 3.2). He felt, although he never came up with much by way of causal explanation, that the continents were made up essentially of sial, floating on the rather denser sea-bed of sima, and they could plough through the sima, as they moved slowly around the earth.

Wegener did not win many converts. Perhaps the most famous of his advocates was the South African Alexander du Toit, who published his own version of Wegener's thesis (du Toit 1937), replacing the single proto-continent of Pangaea with two, Laurasia in the north and Gondwana in the south (this latter is usually, etymologically incorrectly, referred to as 'Gondwana-land'). But generally, reputable geologists brusquely dismissed continental drift as an untenable hypothesis – something to laugh about rather than take seriously.

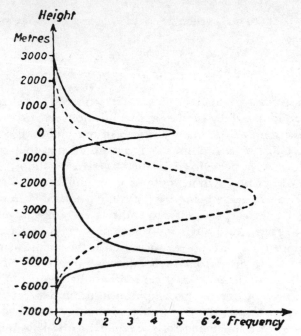

Figure 3.1 The distribution of the earth's elevations.
The solid line represents Alfred Wegener's double-peaked curve of
surface elevations. In Wegener's view the two peaks represent two
fundamental levels, the surfaces of the sial and the sima. The
dashed line is the Gaussian distribution which Wegener would expect if the earth had only one surface level that has been deformed
to create continents and ocean basins. (The diagram was first used
in *Die Entstehung der Kontinente und Ozeane*, second edition, 1920.
This English version is from *The Origin of Continents and Oceans*, translated by J.G.A. Skerl, 1924, p.30)
Source: From Marvin (1973) p.70, reprinted by permission

It was round about 1960 that, like a phoenix from the ashes,
the hypothesis rose again. The key paper seems to have been by
Harry H. Hess, published in 1962 but circulated before that,
entitled 'History of ocean basins', and containing in his own
words 'an essay in geopoetry' (Hess 1962). And yet perhaps talk
of a 'phoenix' is misleading, for it was not really Wegener's
hypothesis that Hess revived. He did not endorse a picture of
sial continents ploughing through sima beds, even though he did
suggest that the continents had shifted around the earth. Rather
Hess suggested that the continents are embedded in sheets (or,

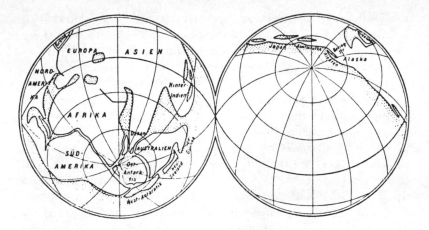

Figure 3.2 Wegener's 'Pangaea' (which first appeared in *Die Entstehung der Kontinente und Ozeana*)
Source: From Marvin (1973) p. 73, reprinted by permission

as they are now known, 'plates') of sea-bed, which sheets slip around the surface of the globe carrying the continents with them. The movement of the plates is somehow a function of the globe's interior, its heat and its chemical composition, and what we have is new material welling up and forming new sea-bed at certain cracks of the earth's surface, the plates thus formed consequently spreading apart, and then at the other cracks of the earth's surface, one plate slipping down beneath another until deep in the earth's interior it is destroyed. (One can therefore see why today's geologists find the term 'continental drift' a little misleading and prefer the term 'plate tectonics', although the earlier term seems stuck in the general imagination.)

Hess's hypothesis was given a dramatic boost in credibility by the ideas of two young Cambridge scientists, Fred Vine and Drummond Matthews (1963). As newly formed earth-surface cools, it becomes magnetized by the earth's field. There was growing evidence that every now and then (irregularly but of the order of half a million years) the earth's magnetic field reverses its direction. This means that, if the earth's surface is growing as Hess suggested, one ought to be able to trace it through geomagnetic reversals. Ideally from a rift (crack) of new growth, one lined up north-south, one would find parallel strips of rock, magnetized in a direction opposite to its neighbours. But, pointed out Vine and Matthews, not only does one find these 'magnetic anomalies', but, as one would expect were the theory of plate tectonics true, one finds that the reversals on one side of a rift are an exact mirror image of those on the other side.

After this, things rapidly fell into place. For instance, new ways (using computers) were found for fitting together continents, thus making the notion of original proto-continents more plausible (Bullard *et al.* 1965). Similarly, some of the more significant geological configurations could now be explained: for instance the Himalayas are the effect of India having drifted up from Gondwanaland and smashed into the under-belly of Asia. But perhaps the most dramatic work occurred in the study of earthquakes. These phenomena occur almost exclusively in certain restricted parts of the world. Through plate tectonics it proved possible to explain their nature, specifically their size, their restricted locations, and their proximity to volcanoes. One could now understand earthquakes as a function of, either the creation of new sea-bed, or more important, the disappearance and destruction of old sea-bed (Sykes 1967, Isacks *et al.* 1968).

So much for the barest outlines of what happened scientifically. Let us turn next to what the geologists think are the philosophical implications of their revolution – for such implications they believe there are.

IS THE REVOLUTION KUHNIAN?

Almost to a man – or a woman – those who have written about the geological revolution think it fits the pattern of a scientific revolution as described and analysed by Thomas Kuhn in his influential *Structure of Scientific Revolutions* (1962).[1] Thus, consider the comments of two of the men who were deeply involved in the revolution. The Canadian geophysicist J. Tuzo Wilson thinks that as in astronomy at the time of Copernicus, we have just had a Kuhnian-type revolution in the earth sciences.

As before, the new beliefs do not invalidate past
observations; the new beliefs depend upon reinterpretations
of geology and geophysics, and they demonstrate the
interdependence of the two disciplines. The acceptance of

continental drift has transformed the earth sciences from a group of rather unimaginative studies based upon pedestrian interpretations of natural phenomena into a unified science that is exciting and dynamic and that holds out the promise of great practical advances for the future. (Wilson 1970, preface)

Similarly Alan Cox of Stanford, a pioneer in work on paleomagnetism, endorses a Kuhnian view of the geological revolution, even preferring to use Kuhn's language.

We have followed Kuhn in using the terms 'scientific revolution' and 'paradigm' to describe plate tectonics and sea-floor spreading, rather than the more traditional 'hypothesis' or 'theory'. In this way a sterile argument was avoided about whether plate tectonics should be described as a hypothesis or a theory. Moreover the development of plate tectonics, although describable in terms of several theories of the history of science, fits the pattern of Kuhn's scientific revolutions surprisingly well. It appears reasonable, therefore, to regard developments in the earth sciences during the past decade as the emergence of a major new scientific paradigm. (Cox 1973, p. 5)

Of the two histories of the revolution that I have read, the one author states:

It seems to me, however, that Kuhn has highlighted major features of science with a most illuminating conceptual model and has been perceptive in challenging the conventional view of cumulative progress. The earth sciences do indeed appear to have undergone a revolution in the Kuhnian sense and we should not be misled by the fact that, viewed in detail, the picture may appear somewhat blurred at the edges. (Hallam 1973, p. 108)

The other author states:

The story of continental drift as a geologic concept, with its slow, tentative beginnings and violent controversy, followed by the spectacular bandwagon effect which has swept up the majority of earth scientists, bears out in dramatic fashion a thesis developed by Thomas S. Kuhn in his book *The Structure of Scientific Revolutions*, first published in 1962. (Marvin 1973, p. 189)

Clearly, the accepted paradigm for analyses of the geological revolution is Kuhnian.

Now, as is well known, since the *Structure of Scientific Revolutions* first appeared, philosophers of science have made something of a pastime of criticizing it. Indeed, not a few Ph.D. theses (of which I confess mine was one) have been brought to successful completion on the basis of subtle and not-so-subtle assaults on Kuhn. Hence, as a philosopher one's initial reaction to all of this endorsement of Kuhn by geologists and their historians might be to dismiss it as naïve, and to turn at once to channels one finds more fruitful and trustworthy. I shall not do this. Apart from the fact that I have not yet found any general account of scientific change less flawed than Kuhn's (and most are a good deal more flawed) such a move smacks of the arrogance that we philosophers of science are ever-ready to find in the attitudes of real scientists towards us. Moreover, one suspects that if so many people knowledgeable about the geological revolution endorse Kuhn, there must be something in his writings which reflects the spirit of recent geology.

I shall therefore take very seriously the claim that the geological revolution was a Kuhnian revolution, and shall look carefully at it. It is true that I shall have some critical things to say – indeed I shall go as far as to argue that in important respects the revolution was not Kuhnian – but I must emphasize strongly that my primary aim will not be yet one more refutation of the *Structure of Scientific Revolutions*. Rather my hope is to use Kuhn's stimulating ideas (and if they are not that, why do we keep talking about them?) as a means towards showing what might be a more adequate analysis of the geological revolution. What follows should therefore be seen in more of a positive than a negative light.

KUHN'S THESIS ABOUT SCIENTIFIC CHANGE

A major difficulty facing anyone who would comment on Kuhn's ideas is that there is a significant gap between what Kuhn thought he said in his *Structure of Scientific Revolutions*, and what everyone else read him as having said. But, assuming that it was the public Kuhn to whom our geologists were responding, the following claims seem fairly central to the Kuhnian analysis of a scientific revolution.[2]

First, there is what one might call the *sociological* aspect of a scientific revolution. Practitioners in a particular area of science are weaned from one theory, or 'paradigm', or *Weltanschauung*, to another. Usually the people involved in making the breakthrough to the new paradigm are fairly young: sufficiently experienced to know what is wrong with the old way of doing things, but not sufficiently set or full of achievement to be emotionally committed to the past. Often the older scientists of the discipline are unable to make the switch; they feel strong hostility to the new paradigm and its supporters; and matters are only resolved as the old-timers die off (Plank's Principle). Associated with a revolution one gets all kinds of scientific infighting: people try to control the journals and what gets published in them; textbooks are rewritten *à la 1984*, with the new paradigm right at the beginning and blunt suggestions that only a fool could fail to accept it (or rather no suggestion at all that anyone might fail to accept it); people with the right beliefs get the right jobs.

Second, we get the *psychological* level to a revolution. People in a new paradigm see things differently from those in an old paradigm. Kuhn is fond of using classic psychological experiments in perception to illustrate his point: now you see a rabbit, now you see a duck; now you see a young woman, now you see an old woman. The flip from one paradigm to another is therefore something of a *Gestalt* experience – now you don't have it, now you do. A conversion experience, happening not gradually but in a flash, would be another way of illustrating things. And as happens with conversion experiences, the newly converted feel inspired and excited: they want to push into the new field and to infect others with their feelings.

Next, we have the *epistemological* dimension to a revolution. When one switches from one paradigm to another, one breaks new ground with respect to methodology and to data. On the one hand, the rules of the game change. What was important or significant in one paradigm is no longer so in the other. What

63

counts as the proper way of doing things, of getting and evaluating evidence, changes. For a pre-Copernican the difference between the inferior and superior planets was interesting but not very significant; for a post-Copernican the difference was crucial and any theory which did not even try to explain it was unacceptable. On the other hand, the data of science gets transformed. Because we no longer see things in the same way, we no longer see the same things. Priestley saw dephlogisticated air; Lavoisier saw oxygen, a new gas.

Fourthly and finally there is the *ontological* dimension to a Kuhnian revolution. At times of revolution, argues Kuhn, in some very real sense it is not simply a question of the world seeming to change for us, but rather the world really does change. Given the only sense of reality that we can understand, reality on either side of a revolution is different. There is no ultimate given against which science responds: our knowledge of the world and the world itself are inextricably bound together.[3]

These various strands run together in Kuhn's work, adding up to a view of science which can best be described as 'relativistic'. There is no ultimate progress in science; no absolute truth towards which science is asymptotically creeping. One can never go back to an old paradigm, but as with styles in music or painting one cannot really say that the present is 'truer' than the past. A great many people have concluded that Kuhn portrays science as an irrational affair. This is not quite accurate. Because the rules of good science change over a revolution, there is no ultimate touchstone of objectivity by which to assess the rationality of a revolution; but this does not mean that a revolution is irrational in the sense of slightly crazy. There are criteria making a revolution sensible: increased problem-solving ability of the new paradigm; simplicity; metaphysical attractiveness; and so forth. Hence, it is perhaps best to say that Kuhn looks upon the movement of science as 'arational'.

SOCIOLOGICAL AND PSYCHOLOGICAL FACTORS IN THE GEOLOGICAL REVOLUTION

How then does the revolution in geology fit with this view of science? Somewhat unfairly my main focus in this paper will be on the third aspect of Kuhn's thesis, the epistemological dimension, because this is the kernel of the philosophy of Kuhn's analysis. I say that my focus is 'somewhat unfair' because a several-year intensive study of the Darwinian revolution has convinced me that sociologically and psychologically Kuhn has much of value to say about scientific change (Ruse 1979a). And I suspect the same is the case here. Hence, although this is supposed to be a 'philosophical' essay, I am certainly not going to ignore these dimensions entirely.

Let us take first the sociology of the geological revolution. There is little doubt that before the 1960s continental drift was generally greeted with contemptuous hostility. One early critic wrote: 'Wegener's hypothesis in general is of the foot-loose type in that it takes considerable liberty with our globe and is less bound by restrictions or tied down by awkward, ugly facts than most of its rival theories' (Chamberlin 1928, p. 87). And one of the historians of the revolution remembers how her colleagues 'laughed heartily' at the hypothesis (Marvin 1973, foreword). Whether there was actual suppression of ideas before the revolution I cannot say, although apparently at least one person ran into trouble publishing pertinent ideas. His hypothesis, one similar to that of Vine and Matthews for explaining magnetic anomalies, was rejected as 'too speculative for publication and more suitable for discussion at a cocktail party' (Cox 1973, p. 226).[4] Also in confirmation of Kuhn is the way that the textbooks have been rewritten since the revolution,[5] and the fact that many of the key figures in the revolution were young: Vine and Matthews, for instance, were just beginning their careers in science (Vine was a graduate student), as also was Cox. Wegener, incidentally, was 30 when he conceived his hypothesis.

However, against Kuhn is the fact that Hess was not particularly young when he had his ideas, nor for that matter was Wilson. Also worth note is the fact that although some geologists still reject continental drift, their numbers are not many, and many who for years had opposed the hypothesis have swung round and accepted it. For instance, James Gilluly, author of a classic text in geology (1974), early editions of which (1959) rejected drift (although in fairness it must be noted that the rejection was sympathetic), has now written that present evidence is 'cumulatively so compelling that the reality of plate tectonics seems about as well demonstrated as anything ever is in geology'

(Gilluly 1971: 648). Whether these contrary sociological facts point to anything of philosophical interest will be considered later. But overall it must be allowed that Kuhn seems sociologically informative. The geological revolution did mainly follow the path Kuhn predicted.

Psychologically speaking Kuhn also scores. We have seen already how liberating and exciting geologists find their new theory, and it seems clear that many of them came to the theory by something very much akin to a conversion experience. One participant, Tanya Atwater has written:

> Sea floor spreading was a wonderful concept because it could explain so much of what we knew, but plate tectonics really set us free and flying. It gave us some firm rules so that we could predict what we should find in unknown places. At Scripps, Bill Menard had been browsing on the origins of fracture zone offsets for a long time and he immediately began trying to make the rules tell him something about them. The distinct bend in the Mendocino set us off on the direction change. At first Bill and I were catching each other at odd moments, scribbling sketches on envelopes and scraps of paper, but we got more and more excited until we began hunting each other up in the morning to compare the previous night's thoughts. It was wonderful working with Bill because he knows the oceans so incredibly well. Whenever we found a new geometrical relationship, he could think for a moment and draw out of his mind some appropriate examples from the real world. The creation of brand new fracture zones by changes in direction of spreading was a prediction that fell straight out of the sketching games; there was no well-documented case. We went ahead and published it – his enthusiastic optimism overriding my trepidation. I was utterly amazed when we got some new lines near the great magnetic bight and the pattern was there, just as predicted. That day I was converted from a person playing a game to a believer. (Atwater 1973, p. 410)

And then:

From the moment the plate concept was introduced, the

geometry of the San Andreas system was an obviously interesting example. The night Dan McKenzie and Bob Parker told me the idea, a bunch of us were drinking beer at the Little Bavaria in La Jolla. Dan sketched it on a napkin. 'Aha!' said I, 'but what about the Mendocino trend?' 'Easy!' and he showed me three plates. As simple as that! The simplicity and power of the geometry of those three plates captured my mind that night and has never let go since.

It is a wondrous thing to have the random facts in one's head suddenly fall into the slots of an orderly framework. It is like an explosion inside. That is what happened to me that night and that is what I often felt happen to me and to others as I was working out (and talking out) the geometry of the western US. I took my ideas to John Crowell one Thanksgiving day. I crept in feeling very self-conscious and embarrassed that I was trying to tell him about land geology starting from ocean geology, using paper and scissors. He was very patient with my long bumbling, but near the end he got terribly excited and I could feel the explosion in his head. He suddenly stopped me and rushed into the other room to show me a map of when and where he had evidence of activity on the San Andreas system. The predicted pattern was all right there. We just stood and stared, stunned. (Atwater 1973, pp. 535-6)

Since so much written about Kuhn in recent years by philosophers has been critical, in his defence I must say that he prepares one for autobiographical recollections like that. Such accounts come as something of a shock to those of us raised on the super-rational picture of science of the logical empiricists. I find it no surprise that it is Kuhn who is a success with working scientists.

METHODOLOGY

We come now to Kuhn's epistemology. (I take it that a necessary condition for the success of Kuhn's ontology is the success of his epistemology, and shall therefore postpone until later any consideration of the former.) Does the recent revolution in geology fit Kuhn's thesis on this score? There are two questions here. First, in the geological revolution did one get a change in the rules and the methods of good geologizing? Second, did the data in some way change, at least inasmuch as we interpret it? Let us take these in turn.

Of course in a trivial way one can say that the rules did change. Before the revolution the rule was: Don't explain using continental drift. After the revolution the rule was: Do explain using continental drift (or plate tectonics). But this is all a bit thin, and certainly not something that would distinguish a Kuhnian thesis from any other. Can we identify some rules beyond these that geologists use and see if they changed? There are, I think, two possibilities: general rules of scientific method, which are not necessarily restricted to geology, and specific rules which may well be peculiar to the science of geology.

As far as general rules are concerned, one might I suppose start with the basic level with things like *modus ponens*, but lest this be thought too stringent (although I do not concede that it is), let us take some of the broadest criteria for what would be considered as good science. These could certainly be incorporated into methodological dicta. An obvious suggestion would be that the scientist ought to strive to produce science that is hypothetico-deductive, that is to say axiomatic with theoretical terms in the premises. Unfortunately, that happy and simple time when one could claim that in the better class sciences the hypothetico-deductive ideal holds, and carry most of one's audience with one, has now passed. Apparently, apart from a few philosophical fossil relics like myself and Mary Williams, no one today believes that the hypothetico-deductive model has any applicability — in physics or anywhere else. We must therefore search on for examples of general rules. Fortunately, however, we do not have to search far, for these days there is another claim about science which is fairly commonly held – positively fashionable in fact – which can certainly furnish a general rule. This is the claim that the best kind of science explains in many different areas from one hypothesis: in other words, that one's science ought to exhibit what William Whewell (1840) called a 'consilience of inductions'.

Now a consilience is certainly something to be found in physics. Newtonian astronomy is the paradigm case of a consilient theory.[6] Starting from his basic axioms, Newton can in turn explain the motions of the planets (for example, Kepler's laws), the motions of terrestrial bodies (for example, Galileo's laws), the tides, the moon. Similarly we find consilience in the better biological theories. Not surprisingly given his influences, we find that Darwin's theory of the *Origin* is consilient, as he explains through natural selection in behaviour, paleontology, bio-geography, taxonomy, embryology, anatomy, and other areas (Ruse 1975b, 1975c). And this consilience is still a distinctive mark of the modern synthetic theory of evolution (Ruse 1973c, 1979b).

What about geology – pre-plate-tectonic geology, that is? Darwin was a geologist before he was a biologist, and as we might expect, some of the work for which he has been most praised – his coral-reef theory (1842) – was consilient. Starting from an overall thesis about the world being in a constant state of elevation and subsidence, Darwin felt that he could explain all sorts of phenomena: how coral reefs were formed, through the gradual sinking of the sea-bed beneath the coral; how it was that barrier reefs, similarly caused by subsidence, always occurred with atolls; why fringing reefs never occurred with other kinds of reef, because they alone were caused by elevation; and other like facts (Ghiselin 1969, Ruse 1979a). In other words, we undoubtedly find a consilience taken as a mark of good science in pre-plate-tectonic geology. Geologists ought to try to be consilient.

Turn now to the new theory or paradigm, if one may so call it without prejudging the issue. Consider the following:

> Certainly the most important factor is that the new global tectonics seem capable of drawing together the observations of seismology and observations of a host of other fields, such as geomagnetism, marine geology, geochemistry, gravity, and various branches of land geology, under a single unifying concept. Such a step is of utmost importance to the earth sciences and will surely mark the beginning of a new era. (Isacks *et al.* 1968, p. 362)

This passage, it must be added, comes in a major paper analysing the application of plate tectonics to seismology (the study of earthquakes). Or go back to Gilluly, after lifetime opposition a convert to the new theory:

So far as I know, no one has yet suggested a model for the generation of plate motion that is acceptable to anyone else. Nevertheless, the arguments from magnetic strips, from the distribution of blue schists and ophiolite belts, from sedimentary volumes, from the mutual relations of volcanic and plutonic belts, ocean ridges and deeps, and from the JOIDES drill cores, are cumulatively so compelling that the reality of plate tectonics seems about as well demonstrated as anything ever is in geology. (Gilluly 1971, p. 648)

So much for new rules or methodology at this level.

Perhaps a better case for the Kuhnian thesis can be made if we restrict ourselves more closely to geology itself. What kind of distinctively geological methodology do we find pre-plate-tectonic geologists using? Do we find any reflection of this in post-plate-tectonic geologists' work? Rightly or wrongly most geologists think that Lyell was the major figure in their history's past, and that it was he, with his commitment to 'uniformitarianism', who set them on the right path. Kuhn, incidentally, seems to agree, for Lyell's *Principles of Geology* (1830-3) is one of his examples of a paradigm-creating work (Kuhn 1962, p. 10). So what peculiarly geological, methodological dicta do we find in the *Principles?*

Historians of Lyell differ strongly, not to say bitterly, over the true interpretation of Lyell's achievements and his real importance for geology (Rudwick 1969, Wilson 1972). But all come together in agreeing that the label 'uniformitarian', first bestowed on Lyell by his friend and critic William Whewell, insensitively masks the several things that Lyell aimed to do, as also does the counter-label 'catastrophism' for Lyell's opponents. Following recent commentators (Rudwick 1972, Mayr 1972a, Ruse 1976a), we can I believe distinguish three things that Lyell was trying to do, or three criteria that guided his geological conduct. First, Lyell was an 'actualist': he wanted to explain past geological phenomena in terms of causes of a kind acting today. He wanted no strange new causes to explain the past. Second, restricting now the use of our term, Lyell was a 'uniformitarian': he wanted to explain past geological phenomena in terms of causes of a *degree* or *intensity* acting today. He wanted no super-forces in the past, although he did allow that sometimes causes really do build up in the effects that they have. Some day Niagara Falls will have eaten its way back to Lake Erie, and what a flood we shall have then! Third, Lyell was committed to a 'steady-state' view of the earth: allowing substantial cyclical fluctuations, Lyell thought the earth in a constant holding pattern, neither building up nor running down.

Lyell's 'grand new theory of climate' illustrates admirably his guide-lines. Faced with fairly strong evidence (fossil palms) that in the past Europe was warmer than it is now, rather than concede that the Earth was on a directional cooling-down, Lyell argued that climate is cyclical, within fairly definite limits. The alterations in climate he explained causally, ultimately in terms of such things as erosion and deposition, subsidence and elevation, which alter the relative shapes of land and sea, and thus set up different ocean currents and so forth. Eventually, this all leads to different climates.

Obviously Lyell invoked his theory of climate to keep within his steady-state guide-lines. But there was more than this. To achieve his ends Lyell did not invent catastrophic causes of kind and intensity unknown – 'without help from a comet' rushing up close to the earth (Lyell 1881, 1, p. 262) – but rather he used forces which we today experience, like the Gulf Stream. In other words, Lyell stayed faithful to his dicta of actualism and uniformitarianism.

Now, if we take these three guides of Lyell and turn to modern geology, what do we find? Perhaps of the three, it is Lyell's steady-statism which we should least expect to find, or rather be least disappointed if we do not find, for Lyell subscribed to steady-statism for what, scientifically speaking, can only be described as very fishy reasons.[7] He thought that inorganic direction leads to organic progression, which in turn leads to evolution and hence to the belittling of man's unique status (perhaps he was right!). Also, Lyell's particular brand of deism liked a world which neither runs up nor runs down, but which ticks along like well-oiled clockwork. However, despite reduced expectations, it would seem that modern geology does give us a little bit of bonus here. Ignoring the question of the beginning (to be picked up in a moment), the picture sketched by plate tectonics is of the huge plates on the earth slowly rising, moving across, and sinking, somehow fuelled by heat, gravity, the

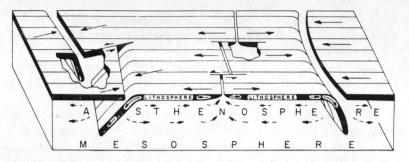

Figure 3.3 Block diagram illustrating schematically the configurations and roles of the lithosphere, asthenosphere, and mesosphere in a version of the new global tectonics in which the lithosphere, a layer of strength, plays a key role. Arrows on lithosphere indicate relative movements of adjoining blocks. Arrows in asthenosphere represent possible compensating flow in response to downward movement of segments of lithosphere. One arc-to-arc transform fault appears at left between oppositely facing zones of convergence (island arcs), two ridge-to-ridge transform faults along ocean ridge at centre, simple arc structure at right.
Source: From Isacks *et al* (1968) p.5857, reprinted by permission

laws of chemistry, etc. (see Figure 3.3). Everything just keeps turning over and over, endlessly. In other words, plate tectonics does rather strike one as being steady-state, even though things change: Africa and South America split apart and India drifted up from Gondwanaland to smash into Asia. However, Lyellian steady-statism never denied this kind of thing. Indeed, it demands a change in the lands and the seas, although this is to be achieved by elevation and subsidence rather than lateral motion. Nor is the finding of steady-statism in plate tectonics to say that today's geologists are deists, and that this influenced their geologizing. But there does seem still to be a commitment to a steady-state world. One might of course argue that things on the present theory could run down, and this is true. But as today's geologists face their problems – the San Andreas fault, the geology of the Gulf of Aden, the Himalayas – they seem to be guided by the rule that everything must be understood as part of an ongoing, constant process.

What about actualism and uniformitarianism (in the restricted sense)? Do modern geologists want to explain in terms of causes of kind and intensity holding today? Clearly they do. Take first the negative side of the rejection of Wegenerian continental drift. To say that it was rejected because it was catastrophic, as one commentator (Marvin 1973) has said, is not really accurate. It was rejected because it failed the tests of actualism and uniformitarianism. Wegener wanted the continents to sail across the earth's surface, rather like a rubber duck sails across the top of the bath water. But the simple fact of the matter is that continents are not rubber ducks and sea-beds are not bath water. As we now understand forces of nature – their kind and intensity – a solid lump of rock stuck in another solid lump of rock simply cannot move. That is that. Hence, because it violated the geological game-rules Wegener's hypothesis was rejected.

Take next the positive side of the argument and look at Hess's classic paper. Hess quite openly concedes what he calls 'the great catastrophe':

> In order not to travel any further into the realm of fantasy than is absolutely necessary I shall hold as closely as possible to a uniformitarian approach; even so, at least one great catastrophe will be required early in the Earth's history. (Hess 1962, p. 23)

In particular, Hess argues that some time, not long after the formation of the solid Earth, there was a single-cell convective overturn which separated out the Earth's core from its mantle and caused the bilateral asymmetry of the Earth's surface (*i.e.*, division into land and water) (see Figure 3.4). After this overturn, because of the separation of core from mantle, such a drastic process was no longer possible, and so the steady-state, plate-tectonic process took over.

However, I am not at all sure that this 'catastrophe' would be so alien and objectionable to a Lyellian. On the one hand, Lyell himself admitted that strange things may have happened when the world was set up:

> He [*i.e.* God] may put an end, as he no doubt gave a beginning, to the present system, at some indeterminate period of time; but we may rest assured that this great catastrophe will not be brought about by the laws now existing and it is not indicated by any thing which we perceive. (Rudwick 1976, p. 148)

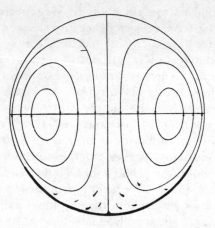

Figure 3.4 Single-cell (toroidal) convective overturn of Earth's interior.
After Vening Meinesz. Continental material extruded over rising
limb but would divide and move to descending limb if convection
continued beyond a half cycle.
Source: From Hess (1962) p. 601, reprinted by permission

This is from a lecture given by Lyell in 1832. On the other
hand, I do not think that Hess even imagined a catastrophe of
the order that Lyell himself was prepared to deny! Hess did not
want different laws of nature, or causes of a kind and intensity
unknown – merely that they come together to effect one unique
major phenomenon: something akin to (albeit much greater
than) Niagara Falls backing all the way up and Lake Erie over-
flowing.

Laying aside this catastrophe, as Hess himself admits, the
commitment to actualism and uniformitarianism (narrow sense)
in his paper is quite remarkably striking. The key fact on which
Hess based his case was that the sedimentation on the sea-bed
is fairly thin, indicating that at current rates of deposition the
sea-bed surface is young (260 million years, for an Earth now
known to be four and a half billion years old). Now to get round
this problem Hess could have assumed causes of a kind and
intensity unknown – an expanding Earth would do the trick,
because then the sediments would be spread more thinly.
However,

I hesitate to accept this easy way out. ... [It] is

philosophically rather unsatisfying, in much the same way as were the older hypotheses of continental drift, in that there is no apparent mechanism within the Earth to cause a sudden increase in the radius of the Earth. (Hess 1962, p. 32)

In other words, because he was firmly committed to actualism and uniformitarianism, Hess felt that from current evidence of sediment deposition he had to conclude that the sea-beds are young (although the evidence is that the continents are old), and therefore he had to suppose a hypothesis to explain this: sea-bed spreading, which is the foundation of plate tectonics.

We find a similar commitment to this methodology in the work of other scientists who have endorsed the new outlook. Those working on geomagnetic reversals, for example, were very concerned to show how causes of a kind and intensity known today can effect various magnetic phenomena in cooling rocks, so that they could then argue that the magnetic anomalies found on the sea-beds can only be explained by spreading, as new bed is created and old bed absorbed (see Cox 1973, especially section IV, 'Geomagnetic reversals: the story on land'). But there is no need to labour the point. The key items of Lyellian geological methodology are obviously as crucial to today's geologists as they were to the earth-scientists of the past.

Finally, to conclude our discussion of methodology, note that even with respect to the most specific of details there is continuity. At least from the time of William Smith, the 'Father of Geology', it has been axiomatic amongst geologists that a good guide to the past is the tracing of various strata (see Laudan1976, for a discussion of Smith's contribution to geology). For instance, if one finds two identical sets of strata separated in some way, one can be fairly certain that they were once together and some cause separated them. But this is precisely the methodological assumption behind one of the most dramatic pieces of evidence for the new geology, namely the exact way that the north-west geology of South America matches the geology of the Gold Coast of Africa (see Figure 3.5). Similarly, for Lyell and Darwin, organic distributions (today's or fossil) were crucial tools in the inference of the geology of the past. Likewise modern geologists turn to organic distributions for help (as indeed, did some of those who argued against Wegenerian continental drift) (Marvin

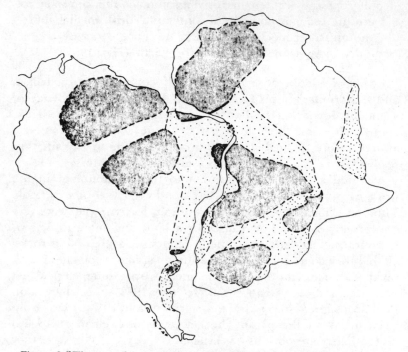

Figure 3.5 The matching of age provinces between Africa and South
America.
Dark gray areas are at least 2,000 million years old; stippled areas,
more than 600 million years old. The heavy line marks the dated
contact that extends from the vicinity of Accra, Ghana, to that of
Sõa Luis, Brazil.
Source: From Marvin (1973) p.159, reprinted by permission

1973). Finally, I must make mention of the frontispiece of Lyell's
Principles, which shows how he used erosion to infer elevation
and subsidence (Figure 3.6). The same inferential machinery
directed towards the same ends is in use today (Figure 3.7).

The conclusion we must surely draw is that, with respect to
methodology, there is strong continuity across the recent geological revolution. At all levels, the guide-lines for good geology are
shared by pre- and post-tectonic geologists. Let us turn next to
the question of the data.

Figure 3.6 Frontispiece to Lyell's *Principles of Geology* (1830-3)

Figure 3.7 Measuring the height of the pre-earthquake upper limit of barnacle growth, or barnacle line, above the present water level at Glacier Island in Prince William Sound.
The sharply defined upper limit of barnacle growth is typical of much of the Prince William Sound area.
Source: From Plafker (1965) p. 1678, reprinted by permission

THE FACTS

Epistemologically speaking, Kuhn argues that the facts of science are 'theory laden'. When we pass through a revolution the facts change in some way because we interpret them and thus 'see' them in a different way. (Ontologically, Kuhn wants to go further than this and argue that the facts really are different.) A favourite example of Kuhn's, one that we might use as a paradigm here, is drawn from the chemical revolution: although Priestley and Lavoisier both discovered oxygen, only Lavoisier saw it as oxygen, as a new gas. Until his death, for Priestley oxygen was dephlogisticated air (Kuhn 1962, p. 117).

With respect to the question of facts, three things stand out in the recent geological revolution. First, it is just not true that everything changed as drastically as from dephlogisticated air to oxygen. People knew of earthquakes and volcanoes before the geological revolution; they know of earthquakes and volcanoes after the revolution – they knew also that they tend to come in certain places, like the Pacific coast of the Americas and not Guelph, Ontario. People knew of major geological phenomena like the San Andreas fault and the Himalayas before the revolution; they know of them after the revolution – they knew also of certain fascinating details, like the fact that high up the Alps there were recent fossil formations. (Lyell's opponents took this as definitive evidence of catastrophes – and indeed, why should they have not done so?) People knew before the revolution that there were some very odd facts of organic geographical distribution, which surely suggested that land-sea boundaries were not always as they are now; people know the same since the revolution.

Of course, one would not deny that the interpretation put on these and like facts (in the sense of explanation) has changed across the revolution. But who would deny this? That was the whole point of the revolution! The question at issue is whether pre-plate-tectonic and post-plate-tectonic geologists experienced radically different things when they experienced (say) an earthquake. Were their perceptions like that of dephlogisticated air and oxygen, with the kind of complete breakdown in communication that that implies? It is hard to say that they were. One does not get the kind of total failure of communication and comprehension that is necessary to give plausibility to a Kuhnian viewpoint. An earthquake is an earthquake is an earthquake. Moreover, one cannot really argue, as Kuhn would have us argue, that pre- and post-revolutionary geologists differed in the status they accorded facts – like the inferior-superior planet distinction before and after Copernicus. Lyell and Darwin knew that earthquakes and volcanoes are important and central to a study of the Earth, as do plate-tectonic geologists. And the same holds true of the San Andreas Fault and the Himalayas, etc.

The second point in this section dealing with facts is that a very distinctive feature of the geological revolution is the enormous amount of new and crucial information that was gathered. If one thinks of the Darwinian revolution, the amount of new material uncovered between 1855 (when virtually no one was an evolutionist) and 1865 (when virtually everyone was) was really not very great. Bates published what he had discovered about (and the explanation of) mimicry in butterflies, *Archeopteryx* was discovered, Prestwich found evidence that man had coexisted with extinct animals, but there was not much more (Ruse 1979a). However, the most distinctive fact about the geological revolution is the large body of new information gathered by such means as co-ordinated worldwide surveys – new information which was absolutely crucial in turning continental drift from an unsupported pipe-dream into a well-confirmed hypothesis.

It was discovered that the sea-bed is covered by only a thin layer of sediment, thus implying its youth. It was discovered that the sea-bed is not absolutely flat, but that running through the oceans are huge mountain ranges, or 'ridges', sometimes with peculiar properties (like a rift or valley running along the summit). It was discovered that the sea-bed does not present a uniformly magnetized face, but that, as in the South Atlantic, there are reversed strips, arranged in a mirror image about a rift (see Figure 3.8).[8] Earthquakes were discovered to have definite patterns – light earthquakes along rifts, light earthquakes in other places (where the plates are supposed to go down) and major deep earthquakes behind these latter light earthquakes (where the plates are supposed to be breaking up) (see Figure 3.9). All of these facts and more were the making of plate tectonics. And yet they had simply not been available to earlier geologists.

Now I am not sure that the point being made here brings much comfort to Kuhn (although I shall suggest later that it may not bring much comfort to others either). His point seems to be that the facts change. My point is that the facts appeared, *for the first time.* One way to defend Kuhn is to say that in the geological revolution we had, not a change of paradigms, but a change from pre-paradigm geology to paradigm geology. But there is a penalty to pay. One now has to allow that until the 1960s geologists were fumbling around like sociologists. And this seems not true. From at least Lyell on, the consensus was that continents and seas stay in place on the globe – although what was continent and what was sea at any particular time was quite another matter.[9]

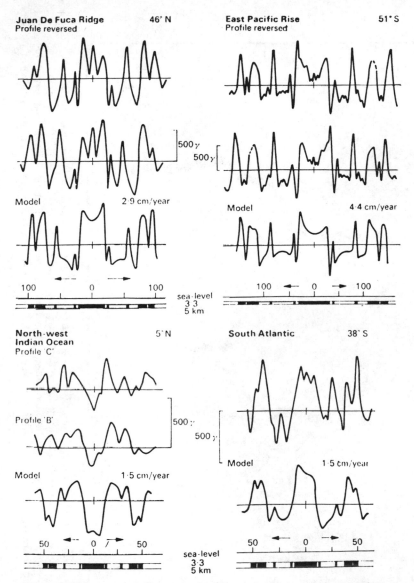

Figure 3.8 Diagram to illustrate the close agreement between observed magnetic profiles and theoretical models based on the Vine-Matthews hypothesis (after Vine 1966, Figs 6-9)
Source: From Hallam (1973) p. 62, reprinted by permission

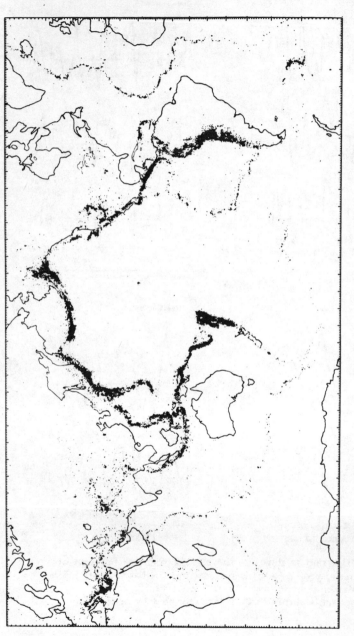

Figure 3.9 Worldwide distribution of all earthquake epicentres for the period 1961 through 1967 as reported by U.S. Coast and Geodetic Survey (after Barazangi and Dorman, 1968).

Note continuous narrow major seismic belts that outline aseismic blocks; very narrow, sometimes steplike pattern of belts of only moderate activity along zones of spreading; broader, very active belts along zones of convergence; diffuse pattern of moderate activity in certain continental zones

Source From Isacks *et al.* (1968) p. 5881, reprinted by permission

The third and final point to be made about the facts is that, notwithstanding what has just been said, there are some things which go further in a Kuhnian direction than anything hitherto mentioned. Take for example the phenomenon that most lay-people think of when modern geology is discussed: the fit of South America with Africa. The pertinent questions to be asked are: What is South America? What is Africa? The most obvious answer is 'areas of land bounded by their oceans'; but in the opinions of geologists matters are not quite that simple. Continents are bounded by areas of shallow sea before the bottom gets really deep: 'shelves'. Why not include the shelves in the continents? It is a little arbitrary to be guided absolutely by the level of the sea. But even if the shelves are included in the continents, there still remains some ambiguity. How deep is 'really deep' and when does a shelf cease to be a shelf?

It cannot be denied that in their matching efforts geologists have chosen different boundaries (Marvin 1973). Indeed, to an outsider there is some suspicion that some first find the best fit and then justify their choice afterwards. But, be this as it may, here at least a case might be made for saying that geologists are not working with raw data, but with 'facts' that have been sifted and influenced by the theory that the geologists hold dear.

Although whether this point is strong enough for Kuhn is another matter. Certainly, there seems no support for Kuhn's strongest, ontological thesis. The facts of continents, shelves, and seas do not really change: it is all a question of where to draw boundaries. Nor even, reverting to epistemology, do we seem to have anything as strong as the oxygen-dephlogisticated air case. I do not think that this fiddling with the coast-lines indicates total inability to see the viewpoint of others (even though one may not accept it). Moreover, the impression one gets is that – although to the lay-person fitting together the continents is the key point in the new geology – to many of the new geologists themselves it is somewhat peripheral. What really counts is the newness of the ocean beds, and magnetic reversals. That one should somehow be able to fit together the continents is a consequence. In short, it hardly seems that here we have something to sway us all back to Kuhnianism.

REVOLUTION OR EVOLUTION?

I hope now it will be agreed that although I have had critical things to say about Kuhn, some understanding of the geological revolution has been gained from the discussion.

We saw that a great many geologists found the geological revolution literally that: revolutionary. They found the new theory of plate tectonics exciting and liberating, and they switched into it rapidly and with gusto. Coupled with this, it seems to have been a fairly general revolution. Although not everyone went along with it (Marvin, 1973, discusses critics) one does not sense the significant and persistent opposition that there was to Darwin's mechanism of natural selection (although there are analogies to the switch in the 1860s of the biological community to evolutionism). These two facts, psychological and sociological, are made much clearer and more understandable by what we have learnt at the level of epistemology. On the one hand there was no call for new methodology, breaking with the ways of the past. Plate tectonicists could be – had to be – as actualistic and uniformitarian as Lyell. Hence, geologists old and new could continue to approach problems in the ways that they had always done. Methodologically there was no revolution. On the other hand, what was crucial about the revolution was not that the facts changed, but that so many new facts came tumbling in. As with methodology, we do not get the strain of being required to reject what has gone before; but at the same time there was so much new information requiring new ideas and theories, that when plate tectonics was proposed geologists leapt forward rather as if they had been converted at a revivalist meeting.

In other words, what I see as the key to the geological revolution was the vast amount of new information gathered, certainly since the war, but perhaps even more concentratedly between (about) 1955 and (about) 1965. Unlike Kuhn we can perfectly well say that the revolution was rational, because there was no essential change in geological methodology – all the standards of good geology remained in place. Unlike Kuhn we can also explain why so many, including the older people, went over to a moving Earth – they were able to do so simply because the rules did not change, nor did all the facts of the past have to be thrown away or recycled. But like Kuhn, although not for the same reasons, we can say why geologists really went through conversion: there were so many new facts – not retreads, but brand new radials.

So, if we are not going to be Kuhnians, what then are we going to be? In recent years a number of philosophers of science have endorsed 'evolutionary' metatheories of scientific change, as opposed to Kuhn's 'revolutionary' metatheory (for example Popper 1972, Toulmin 1972). I do not mean that these philosophers of science deny that what we commonly call 'revolutions' occur in science. Rather, the point is that these significant changes are not abrupt as Kuhn argues, but more gradual and in significant respects analogous to evolution as we encounter and understand it elsewhere. There are however two points which make me initially dubious about the application of these evolutionary metatheories to our revolution, one general and the other specific.

Generally, my worry stems from the fact that although one can of course mean very much what one likes by 'evolution' – certainly everyone else has done so in the past hundred years – if one wants to say something usefully informative, rather than simply reiterating the tautology that scientific change involves change, presumably one wants to draw a significant analogy between scientific evolution and evolution as it is elsewhere most commonly experienced and thoroughly understood, namely in biology. But straight away one runs into a monstrous disanalogy, which makes talk of theory 'evolution' a lot less interesting than one might have expected: the raw data of biological evolution, mutations, are random in the sense that they do not occur when needed nor are they directed towards the adaptive ends of their possessors. However, new innovations in science very much are produced to suit the need – Vine and Matthews, for example, introduced their hypothesis specifically to explain the funny things they found about the magnetism of the sea-floor. A brilliant creative scientist is not akin to a gene going wrong. For this reason I wonder how valuable talk of theory evolution, including geological theory evolution, really is.[10]

My specific worry about an evolutionary interpretation of the geological revolution is that, to me, 'evolution' implies some sort of gradual change. Certainly there have in the past been 'saltationary' evolutionary theories involving jumps – indeed, even today it is thought that something along this line holds in parts of the plant world – but if things get too saltationary then evolution goes out of the window. However, I am not sure that we really get anything of the gradualness required for evolution in the geological revolution. On the one hand, methodologically we do not seem to have much change at all. On the other hand, the change that we do have – much new basic information and new theories to explain it – seems to have come fairly abruptly. This does not seem very evolutionary. Nor does it help much to mention Wegener, suggesting that his speculations point to a gradual change to a moving Earth-surface. Most did not accept Wegener when he hypothesized, and as we have seen, when there was a switch it was not really to Wegener's position or one like it.

I suspect that at this point I am going to be in hot water with the historians – or perhaps hot lava. All who have written historically on the geological revolution have been at pains to show that the plate tectonicists had their predecessors, and that moreover these predecessors were not simply Wegener. Hess, for instance, had a predecessor (and influencer) in the brilliant geologist Arthur Holmes, who was putting forward the idea of ocean-floor spreading as far back as 1931. (See Holmes 1931.) And the whole question of geomagnetic reversals and their record in the rocks, so crucial to modern geology, goes back at least to the work of the Japanese paleomagnetist of the 1920s, Motonari Matuyama (1929). But, whilst I cannot deny these predecessors I am rather inclined to play them down somewhat, at least in the context of the point I am making. The matter at issue is whether we see a gradual change in the opinions of the members of the geological community spread over a number of years; not whether we can dig up the occasional precursor. I suggest we do not see such a change. The common tale is that right into the 1960s people laughed at continental drift. Then they stopped laughing, and switched.[11] Moreover, I would point out – at the risk of alienating my historical colleagues yet further – that on this matter the accounts of historians are highly suspect. Almost by definition, historians (even including Kuhn when he wears that hat!) are bound to be evolutionists, imposing an evolutionary interpretation on history. (See Kuhn 1957. Note the sub-title of Marvin's 1973 book!) No one wants a description of forty-five identical arguments opposing Wegener. The historian's job is to dig out the brilliant anticipation of Hess, even though it was the work of one man and went unheeded. If this is not the historian's task, then why all the interest in Mendel? In short, I stand by my claim that the geological revolution was not so very evolutionary.

By this time it might be felt that I am arguing that there is something very mysterious about the geological revolution. This is not really so. What I am arguing is that certain current metatheories of scientific change are not really applicable, although this is not to deny that, as I hope this paper shows, the metatheories can stimulate one towards a more adequate analysis. My claim is that the revolution in geology was less abrupt than a Kuhnian would have it; more so than an evolutionist would have it. I believe that there was a continuity of methodology across the revolution – a very strong and crucial continuity – and for this reason one can certainly speak of the revolution as being rational.[12] The set of canons of good geological science developed before plate tectonics drove geologists to accept the theory, despite in many cases a long lifetime's opposition to continental drift. But the change was rather abrupt, and the key to this lies in the fact that the crucial cause behind the revolution was the wealth of new information – knowledge of the thinness of the sedimentation of the oceans; evidence of the varied state of the sea-beds, particularly of the rifts; the finding of geomagnetic reversals and the uncanny way they get mirrored across rifts; the exact plotting of earthquakes and of their intensities; the fit of geological strata between Africa and South America; and much more. It was not, as Kuhn would have it, that old facts vanished and new facts appeared in their stead; rather, new facts appeared where none had existed hitherto, and their cumulative weight was to make geologists plate tectonicists.

CONCLUSION

Apart from historical/scientific inaccuracies, I am very much aware of this paper's philosophical inadequacies. I hope that I have highlighted a few of the salient features of the geological revolution and perhaps have pointed in the direction of finding an acceptable philosophical analysis. But I have certainly done no more. I have provided no analysis of the exact way in which geologists think their theorizing fits and explains the facts. Is the hypothetico-deductive ideal really inadequate here, and if so, what substitute should one offer? Without some sort of answer to these questions I cannot see that analysis of the geological revolution can go much further, for one must see how it is that geologists think that their new theory adequately explains all of the new information. Also missing in this paper is any real discussion of what some philosophers have called the 'causal-phenomenal' dichotomy in science. Where do today's geologists stand on the ultimate causes of what makes the Earth's surface move around, how much room exists for different opinions, and how do causal speculations relate to lower-level parts of modern geological theory? Is modern geology irreduc- ibly historical in a way that physics is not, or does the revolution in geology take us one step closer to the overall unification of the sciences, as some have argued has been the case with the revolution in genetics of the 1950s (Schaffner 1967, 1969, 1976; Ruse 1973c, 1976c; but see Hull 1972, 1973b, 1974, 1976b)?

I could go on asking questions almost indefinitely. But enough is enough. Hopefully I have stimulated some members of my audience to take up some of these problems. Unless of course they feel more stimulated to show what little I have done must be done all over again, properly![13]

NOTES

1 The one exception is David Kitts (1974). What I shall have to say, I believe, complements Kitts's views, rather than contradicts them.

2 Since the geologists were reacting to the Kuhn of the *Structure of Scientific Revolutions* (1962), I shall not feel obliged to take account of Kuhn's later work.

3 One of Kuhn's more sensitive interpreters, Fred Suppe (1974), has suggested that really such an extreme ontological thesis should not be ascribed to Kuhn. I think that it fairly can, but as will be seen, for the purpose of this discussion not much hangs on the point.

4 The author was L.W. Morely, a Canadian. The explanation which naturally comes to mind to anyone living North of the Border was that the rejection was a direct function of Morley's nationality. However, although this reason has a certain satisfyingly masochistic flavour, and perhaps even has some truth – Vine and

Matthews had their paper accepted by *Nature* in 1963 – it must in fairness be noted that T.J. Wilson, a leader of the new geologists, is Canadian.

5 Compare the second (1959) and fourth (1974) editions of J. Gilluly *et al.*'s classic textbook, *Principles of Geology*.

6 I take it that no one would deny that Newton's theory was a good theory, even though there is today some question about its truth. Certainly we should be dubious of following Whewell and, without qualification, allowing that a consilience is a mark of necessary truth. See Laudan (1971).

7 This is not to say that Lyell was thereby any less a scientist, or any less a good scientist. Although most like to pretend that it is not so, many scientists are driven by non-scientific reasons, and it is not easy to think of anyone at the time of Lyell writing on geology in Britain who did not have some theological axe to grind.

8 This apparently was very influential in changing people's minds.

9 Laudan (1977: 134-5) has some interesting comments on the extent to which, if at all, Lyell's work could be said to have created a geological paradigm. I suspect however Laudan would agree that there is something a little odd about claiming that until 1965 geology was entirely pre-paradigmatic in a Kuhnian sense.

10 Popper tries to get around this difficulty by making biological evolution quasi-directed. See Ruse (1977b) for criticisms of this ploy.

11 A revealing, but I am sure typical, autobiographical fragment is to be found in Stephen Gould's recent book *Ever Since Darwin* (1977). He tells how, when a student at Columbia, his professor primed an audience to sneer at a drifter. As in the parable of the prodigal son, the most important part comes right at the end: that self-same professor was converted and spent the last two years of his life actively promoting the cause he had always opposed.

12 I suspect that what I am saying may bring joy to the heart of Larry Laudan. He emphasizes the distinction between a theory and a 'research tradition': 'a set of general assumptions about the entities and processes in a domain of study, and about the appropriate methods to be used for investigating the problems and constructing the theories in that domain' (Laudan 1977: 81). In Laudan's terminology, my point is that the geological revolution was rational because, although the theory changed, the research tradition remained the same.

13 Since completing this paper I have been sent a major contribution to the philosophical study of the geological revolution by Henry Frankel (1979).

Part II

CONTEMPORARY ISSUES

As many people have noted, *On the Origin of Species* was not very well titled. Darwin certainly discusses species, but their distinctive properties and ontological standings are not his primary focus. However, biologists have been making up for this ever since, and much has been said and argued about the nature, status, and origin of those basic divisions of the organic world, between *Homo sapiens, Passer domesticus, Drosophila melanogaster*, and all of the other like entities. In recent years, indeed, discussion of these topics has reached even greater dimensions, thanks to a controversial new thesis. It is claimed that the 'species problem' – What are species and why do they (of taxa) uniquely seem to be real? – has been predicated hitherto on a simple but fundamental philosophical mistake. Truly, appearances to the contrary, species are not groups of organisms but are, rather, whole integrated *individuals*. Organisms evolved and so, at a higher level, did species. Hence, species (like organisms) are things with internal integration and species (like organisms) are real because they have spatio-temporal existence.

What makes this claim both exciting and pertinent to my concerns here is that enthusiasts for this species-as-individuals (s-a-i) thesis claim that it can be evaluated only against a background of established biological thought. Agreeing with this latter point, in the first essay of the section I argue that for the Darwinian the traditional conception of species as groups or classes is to be preferred. However, I try to show also how with such a species-as-group perspective it is still possible to preserve our conviction that species uniquely reflect objective divisions in the organic world. It comes as no surprise to me that the way in which this is to be done rests on the classificatory equivalent of the methodological principle which secures the truth of Darwin's theory: a consilience of inductions.

Part of my argument in this first essay depends on showing how uneasily the s-a-i thesis sits with the Darwinian commitment to individual selection. Given that the originators of this thesis are Michael Ghiselin and David Hull, and given also their (already-noted) sensitivity to Darwin's own work, I believe there are significant tensions in their own belief-sets. Nevertheless, as you will infer from my next essay, on the new controversial theory of 'punctuated equilibria', which sees a discontinuous fossil record supposedly reflecting a jerky evolutionary process,

I am inclined to think now that my negative comments in the species paper are a little strong. If you accept a more holistic view of the evolutionary process than do I, then the s-a-i thesis sits much more comfortably.

But, should you accept such a view? Punctuated equilibria supporters think that you should, and they back their case by a range of arguments from paleontology to metaphysics. At one point, as reflected in my species essay, I thought punctuated equilibria theory was just plain wrong. Now, although it is still not a position towards which I am drawn, I think that it represents an older, even non-evolutionary, alternative perspective on the organic world to that of Darwinism. And in this second essay, I try to tease out the antecedents to this alternative perspective, and (reflecting themes introduced earlier) I suggest that we have before us a paradigm difference in some, although not all, of Kuhn's senses.

We have now a dilemma. If you agree with me here about the pertinence of Kuhnian categories, and if you agree also with the corollary that decisions between Darwinism and its rivals are not to be made simply on matters of fact, how then is one to draw conclusions? As I argue in this second essay, a crucial component of Darwinism, one to which natural selection is supposed to speak and which is missing (or, at least, downplayed) in the tradition represented by punctuated equilibria theory, is the commitment to adaptationism – to the belief that the organic world functions and that the parts of organisms contribute to such functioning. In recent years, much has been written on this claim, both for and against it. Elsewhere, I myself have had strong things to say in favour of adaptationism, and do in fact touch on some of the pertinent points in my third essay.

But, rehearsing yet one more time the defence of adaptationism is not my main concern at this point. Rather, I explore the implications for the Darwinian of the belief in adaptationism on the logical structure of his or her thought. In particular, I argue that (like the punctuated equilibria theorist) in respects the Darwinian is part of an older, non-evolutionary tradition. This is a tradition of teleology, of explaining parts in terms of ends, as occurred in the natural theology which was so influential on Darwin himself. Such might seem no great recommendation for Darwinism. Surely the idea of nature working towards goals, perhaps through the medium of vital forces, is the last thing to be found in any self-respecting science? Yet, there is no need to worry here on that score. In fact, one of the most crucial things about Darwinism is how it eschews teleology of this ilk. Mistaken and ignorant though Darwin undoubtedly was about the true causes of heredity, he was adamant always that the new variations on which selection acts – the 'raw stuff' of evolution – appear 'randomly', without respect to the needs of their possessors. There is no end-directed process of this kind in Darwinism.

Nevertheless, I do argue that there is a teleology in Darwinism. But, as I try to show, not only is such teleology non-pernicious, but given the distinctive nature of the organic world, it is an absolutely crucial aid to the forward progress of our scientific understanding. Remember, one of the most important things about a paradigm is that it keeps you moving onwards.

Chapter Four

BIOLOGICAL SPECIES

Natural kinds, individuals, or what?

The status of biological species continues to attract attention and controversy. (See, for instance, Eldredge and Cracraft 1980, Gould 1979a, Grant 1981, Levin 1979, Mayr 1982, Wiley 1978, 1980, Splitter 1982, Mishler and Donoghue 1982, Holsinger 1984, Kitcher 1984, Eldredge 1985b.) There is a strong feeling among biologists, at least there is a strong feeling among most zoologists and somewhat less of one among botanists (a difference to be discussed later), that species are somehow different from the other groupings of organisms we find (or make) in nature. Species, like *Drosophila melanogaster* or *Canis lupus*, are thought to be 'natural', in some way objective or existing independently of the classifier. In this, species differ from the groups (taxa) found at other ranks, for instance that of the genus. The classifier's own thoughts and aims have a much greater role to play in the delimiting of members of these other groups.

But wherein lies the naturalness of species? With the coming of evolutionary theory, traditional answers seem less than adequate. Recently, in a brilliantly innovative move, the biologist Michael Ghiselin (1966, 1969, 1974a, 1974b, 1981, 1987) supported by the philosopher David Hull (1975, 1976a, 1978b, 1979, 1981) has argued that evolution shows us to have misconstrued the nature of species. They are not groups or classes of organisms, like hockey players on a team. Rather, they are integrated *individuals*, with organisms having the relationship to their species of part to whole rather than member to class. And thus properly seen, argue Ghiselin and Hull, the specialness or naturalness of species is self-evident. Species are natural or real in the way any biological individual is natural or real.

In this essay, I argue that, stimulating though the species-as-individuals (s-a-i) thesis may be, it runs counter to much accepted biological thinking, as well as to logic. We must rely on more traditional conceptual tools to establish the naturalness of species. But, with some exceptions and qualifications, this can be done, and the very exceptions and qualifications themselves establish the correctness of the overall approach.

BIOLOGISTS ON SPECIES

Let us start with what biologists have to say about species. The most interesting fact is that the category of species can be and is characterized in so many different ways, with corresponding ways of defining particular taxa names. Here, four major species concepts will suffice. (See Mayr 1982, and Grant 1981a, for recent discussions of the multiplicity of species concepts.)

First, we have the most obvious and intuitive concept of all. We find the organic world broken up into groups of similar looking organisms, with gaps between the groups. The concept thus refers to some notion of overall similarity of appearance possessed by organisms within species taxa. In Charles Darwin's words, a species is 'a set of individuals closely resembling each other' (Darwin 1859, p. 52).

Physical nature or *morphology* is the key to this species concept, and it is therefore invoked when one deals with particular species taxa. To be a member of *Homo sapiens* you must be relatively hairless, capable of walking upright, with a large brain, and so forth. In fact, today it is recognized that invariably there is diversity, even within such morphologically delimited groups, so biologists frequently use polytypic or polythetic definitions: lists of features, a combination of which is sufficient for species membership but no one of which is necessary (Beckner 1959, Simpson 1961, Hull 1965).

Next, we have a concept which has, perhaps a little strongly, appropriated unto itself the title of *biological* species concept. This refers to breeding, or the lack of it. One well-known formulation, due to Ernst Mayr, states that species are 'groups of actually or potentially interbreeding natural populations which are reproductively isolated from other such groups'(Mayr 1942). You do not normally find taxa name definitions using this concept, but I do not see in principle why not. Specify some individual, say Brigham Young, as your reference point, and then members of the same taxon are potential or actual interbreeders, with some obvious qualifications to take account of sex, and so forth. (See Mayr 1982 for his attempts to give a more refined version of the concept.)

Third, we have a concept which deliberately refers to *evolution*. In the words of the paleontologist G. G. Simpson: 'An evolutionary species is a lineage (an ancestral-descendant sequence of populations) evolving separately from others and with its own unitary evolutionary role and tendencies' (Simpson 1961: 153). A taxon name would get a related definition. If we suppose that humans first appeared about half a million years ago, *Homo sapiens* is the name for the group which has descended from the original organisms. (This certainly seems to be the kind of definition that paleoanthropologists have in mind. See Johanson and Edey 1981, Johanson and White 1979.)

Fourth and finally, we have a concept which does for the world of genes, what the first concept did for morphology. The category concept refers to overall *genetic* similarity clusterings, such clusterings being separated from others by gaps. Turning to Mayr again: 'When an evolutionary taxonomist speaks of the relationship of various taxa, he is quite right in thinking in terms of genetic similarity, rather than in terms of genealogy' (Mayr 1969). A particular species name would be defined in terms of genes held in common (together perhaps with information about chromosomes, structure, etc.). For obvious reasons, you do not often see definitions of this ilk, but they do exist. With increasingly sophisticated methods of analysing genomes, their similarities and differences, we might expect to see more such definitions.

There are other concepts which could be and sometimes are invoked, for instance concepts based on ecology (Van Valen 1976). But we have enough for our purposes. The question to be asked now is why a taxon which falls into a category characterized in one of the above ways should be thought natural or real in some sense. Since this is a question *about* science, rather than *within* it, we turn to philosophy for guidance.

NATURAL KINDS

Traditionally, philosophers have treated the status of species as being part and parcel of a larger question about the reality of 'natural kinds'. Why do we think the whole physical world to be divided into different sorts of things: gold, water, stars, as well as *Homo sapiens* and *Drosophila melanogaster*?

Roughly speaking, there have been two main answers to this question (Ayers 1981).[1] Credit for the first is given to Aristotle. He argued that the world – at least, the world of scientific enquiry – is made up of substances. Any particular substance, like a sample of gold or an individual man, results from the interaction between the substance's underlying matter and its form. This latter gives a substance its nature or essence. Objects of the same kind, like two human beings, are the same because they have the same form, which is embedded in different samples of matter. Substances have their form essentially, that is to say, one cannot be a substance of a particular kind without having the required form.

Crucial to the Aristotelian position is the distinction between a 'real definition' and a 'nominal definition'. The former enables you to define a natural kind name, including the name of an organic species, in terms of attributes which stem necessarily from the very essence of a substance (Aristotle called these attributes 'properties'). Thus, in the case of *Homo sapiens* the essence involves the notion of rationality. Unpacking, we get such properties as the power of speech. A real definition would consequently refer to this power. However, not all attributes of an individual stem from the essence. There are features which are possessed 'accidentally'. Although these features are non-essential, it might nevertheless be possible to distinguish a natural kind using only accidents. In the case of humans, both bipediality and featherlessness are accidents, and yet it so happens that the set of featherless bipeds is one and the same as the set of rational animals. Any characterization in terms of accidents yields a 'nominal definition'.

The great rival to Aristotle's analysis came in the seventeenth century, from the pen of John Locke. He argued that reality lies in the underlying particles which go to make up any particular substance. Locke himself was, in fact, doubtful that we could ever truly know these basic units. But any real definition would have to make reference to these building blocks, specifically to their shape, structure, motion, composition, etc. Surface definitions are simply marks of the structure beneath. Consequently, any surface definition could never be more than nominal. For Locke, a definition of humans in terms of rationality has no more and no less status than a definition in terms of bipediality.

But what of the underlying real structure? Even here Locke wanted to deny the absoluteness of Aristotelian essences. Shapes can change, taking a substance from one kind to another. And this is not to mention borderline cases:

> There are Animals so near of kind both to Birds and Beasts, that they are in the Middle between both.... There are some Brutes, that seem to have as much Knowledge and Reason, as some that are called Men ... and so on till we come to the lowest and the most inorganical parts of Matter, we shall find everywhere, that the several Species are linked together, and differ but in almost insensible degrees. (Locke 1975, III, vi, p. 12)

Hence, for Locke even definition in terms of reality ultimately involves a conscious decision to divide. A Lockean definition is therefore never more than what an Aristotelian would label 'nominal'. Any difference between men and changelings 'is only known to us, by their agreement, or disagreement with the complex *idea* that the man *Man* stands for' (Locke 1975, III, vi, p. 39).

In short, whereas for Aristotle natural kinds are ontological entities, for Locke they are at best epistemological concepts. It is the objective approach, versus the subjective approach. The approach which *finds* natural kinds, and the approach which *makes* them.

BUT ARE BIOLOGICAL SPECIES REALLY NATURAL KINDS?

Let us now try to put biology and philosophy together. Does either Aristotle or Locke capture the biologist's sense that species are real or natural?

The simple answer is that, as they stand, neither does. Take

Aristotle. He would argue that species are real, because they are natural kinds. But how then could one get a real definition? The morphological approach to taxa will not do, because if there is one thing that modern biology teaches, it is that evolution promotes morphological diversity. Species members are not all the same. Hence the need for polytypic definitions. But for something to be an Aristotelian property (as opposed to an accident), it must be possessed by every member of the kind, and distinguishing the group from others. Polytypic definitions are not enough. Hence, morphology will not do. (Mayr 1963, Dobzhansky 1970, Dobzhansky *et al.* 1977, discuss variation within species.)

The same considerations apply to genetic features. Evolution promotes genetic diversity (Lewontin 1974). And, similarly, the features relied on by other approaches to species fail the Aristotelian. Is the sterile worker ant even a potential inter-breeder, and would one really want to say that an entirely artificially produced fruit-fly could never really be in *Drosophila melanogaster?*

In any case, after Darwin, strict Aristotelianism simply will not work (Hull 1965, Mayr 1969, 1982). Evolution says that you can take virtually any property you like, and if you go back (or forwards) enough in time then ancestors (descendants) did not (will not) have it. But this is just what Aristotle cannot handle. The whole point about a natural kind is that its properties exist in perpetuity, like mathematical objects. And clearly such properties have to be passed on by ancestors or to descendants. Evolution denies this.

What about Locke? Initially, things seem very much more promising. Morphological criteria give you nominal definitions of species names. Alternatively, if you favour phylogenies, you use relationships of descent – something which Locke himself, incidentally, was not enamoured of. ('[M]ust I go to the Indies to see the Sire and Dam of the one, and the Plant from which the seed was gather'd, that produced the other, to know whether this be a Tiger or that Tea?' (Locke 1975, III, vi, p. 23).) Then, genetic criteria give you the closest things you can get to real definitions. And contrary to Locke's own doubts, we can know quite a bit about these. Nary an issue of *Science* or *Nature* appears without fresh details of the genetic structure of some organism.

Versions of a neo-Lockean proposal are often found in the literature (for instance, Mayr 1963, 1969). Unfortunately, you purchas your solution to the species problem at too high a price. You have to relinquish claims to the ultimate objectivity or reality of species. A Lockean natural kind is essentially subjective or arbitrary. And that is just what you do not want to concede, when it comes to species. In some sense, species are real!

We are caught in a dilemma. Evolution refutes Aristotelianism; but Lockeanism is inadequate. According to Ghiselin and Hull, and an increasingly large number of supporters, we must go back to biology. When we do this, we see that Aristotle and Locke share a false premise. Reject this premise, and hope rises for a solution to the species problem. (Sympathisers with Ghiselin and Hull include Mayr 1976, Wiley 1978, Rosenberg 1980, Sober 1980, Beatty 1982, Splitter 1982, Eldredge and Cracraft 1980. Intimations of the Ghiselin/Hull approach are to be found in Theodosius Dobzhansky's classic *Genetics and the Origin of Species* (1951).)

SPECIES AS INDIVIDUALS

Aristotle and Locke agree that species are natural kinds. The taxon *Homo sapiens* is a class, with individual humans like Michael Ruse and Charles Darwin as members. I qualify for membership in the class *Homo sapiens*, because I possess certain attributes, whatever they may be. So does Charles Darwin. My dog Spencer does not have these properties, and thus does not qualify. He has his own species, *Canis familiaris*.

Ghiselin and Hull argue that species are not natural kinds at all: they are not classes with members. Rather, species are *individuals*, just as particular organisms are individuals. Hence, just as the relationship between my arm and myself is one of part to whole, rather than member to class, so my relationship to the species *Homo sapiens* is one of part to whole. I, and Charles Darwin, are parts of the human species, just as Spencer is part of the species *Canis familiaris*.

The reformers argue that, once we see species in the true light, all of the problems about species start to fade. Of course, species are real. No one doubts the reality of Michael Ruse, or of

Spencer. They are individuals – real things – part of the furniture of our world; so are species.

What kind of claim is the s-a-i thesis? It is not solely or even primarily an empirical claim. 'Look! There's an individual!' Rather, it is more of a conceptual claim, whose plausibility must be argued for. Consider a chess-board. You can think of this as an individual, made up of sixty-four parts, or as a class of sixty-four squares. It depends on your perspective as to which makes more sense – are you making chess-boards, or are you teaching someone the rules of chess? *The crux of the s-a-i thesis, therefore, is whether modern evolutionary biology inclines one to treat species more as individuals, or more as classes, as natural kinds.*

There is of course the initial question as to what precisely one means by an 'individual'. Ghiselin and Hull point to the fact that, whatever else an individual may or may not be, we recognize organisms as paradigm examples of individuals. Organisms are not just diffuse, artificially created entities. They are integrated beings. They have internal organization. Hence, if we can show that in important respects species are like organisms, we can reasonably say that species are individuals.

But, claim Ghiselin and Hull, from an evolutionary perspective species have the very marks of individuality that organisms have. Just like organisms, species come into being, exist for a period in space and time, and then go. And, they have at least some sense of organization. As the leading evolutionist Ernst Mayr said (before he himself was converted to the s-a-i thesis): '*Species are the real units of evolution*, they are the entities which specialize, which become adapted, or which shift their adaptation' (Hull 1976a, p. 183, quoting Mayr 1969).

That evolutionary theory treats species as individuals becomes clear when we look at their uniqueness. Adolf Hitler was an individual – as such (unlike his diaries) he came uniquely and went uniquely. A copy of Adolf Hitler is not he. It cannot be. Similarly, we have this uniqueness for species. 'If a species evolved which was identical to a species of extinct pterodactyl save origin, it would still be a new, distinct species' (Hull 1978b, p. 349). If you are a species, you simply cannot be born again, any more than Adolf Hitler can be.

What about change? Organisms can undergo major change, and still be the same organism. What counts is continuity. The limits of an organism are birth and death. The same is true of species. 'There is no limit to the genetic change that can take place in a species or population before it becomes extinct or speciates' (Hull 1976a, p. 182). In fact, just as in an organism, so long as the continuity persists, we have the same species.

Finally, let us mention one revealing point. Biologists take one specimen from a species, using it as the marker. The species name (which, as with all individuals, is a proper name) is attached to this marker – the type specimen – by an act which is akin to baptism. This specimen does not have to be a 'typical' member of the species, whatever that might mean. The type is part of the whole, not a member in the class. 'The fact that any specimen, no matter how atypical, can function as the type specimen makes no sense on the class interpretation; it makes admirably good sense if species are interpreted as individuals' (Hull 1976a, p. 175). All in all, whatever 'common sense' may say, modern evolutionary biology demands that species be regarded as individuals. Hence, the naturalness of species.

WHY SPECIES ARE NOT INDIVIDUALS: BIOLOGICAL OBJECTIONS

Ingenious though it is, the Ghiselin/Hull attempt to slice through the Gordian knot constraining the species problem fails. There are several significant reasons why species cannot properly be considered as individuals.[2]

First, look at matters at the most basic biological level. We think organisms are individuals because the parts are all joined together. Charles Darwin's head was joined to Charles Darwin's trunk. But, in the case of species, this is not so. Charles Darwin was never linked up to Thomas Henry Huxley. Of course, you might object that although Darwin's head was never linked directly to his feet, they were linked indirectly through intermediate parts. Analogously, as evolutionists presumably we believe that Darwin and Huxley were linked by actual physical entities (namely, the succession of humans back to their shared ancestors). But this objection fails, for the point is that these links have now been broken and lost. If (gruesome thought!) Darwin's head were physically severed from his feet, we would certainly have no biological individual.

Yet, with justice, Ghiselin and Hull will respond that these speculations are beside the point. The required condition for individuality is not mere spatio-temporal contiguity. It is rather some sort of internal integration or organization. Because of such internal organization, the USA is one country, even though Alaska and Hawaii do not touch any of the other states. The fifty states work together, in a way that (say) the forty-eight mainland states together with Ontario and Quebec do not. Analogously, Charles Darwin is an individual, not because of spatio-temporal contiguity, but because his parts are organized, working *together*. The same is true of other biological entities, even those which break into parts at some points in their life cycles, like slime molds. And the same is true of species. They have an integrating organization, with the parts contributing to the whole.[3]

But this will not do, at least not in the light of much modern thought about the working of evolution. First and foremost, thinking of a species as an integrated individual goes flatly in the face of the way in which the major evolutionary mechanism of natural selection is generally regarded today. Selection leads to adaptations, features which help organisms in life's struggles for survival and reproduction. But who precisely benefits from adaptations? Is it the possessors alone, or do others benefit? In short, at what level of biological organization does selection work? Is it between individuals, benefiting individuals, or is it between higher entities like species, benefiting species taken as a whole?

Until recently, most people casually assumed that selection could work at virtually every level of biological organization. In particular, there could be selection between groups of organisms, including between species. The units of selection, in vital respects, were species. As Mayr said: 'Species are ... the entities ... which become adapted' (1969). However, majority opinion today is that selection just does not work in this way. As Charles Darwin himself argued, ardently, selection must work chiefly if not exclusively at the level of the individual organism. 'Group selection' at the level of the species does not work. A species is not adapted. An organism (or, at most, a limited number of organisms) is. Any species effects are just epiphenomena on individual effects, or at most, on population effects. (See Brandon and Burian (1984), for a review of this topic, and Ruse (1980a) for Darwin's views on the subject.)

If this is all so, then there is something very odd indeed about speaking of a species as an individual. It is very far from being an integrated unit like an organism. The individual organisms of a species are all working for their own benefits, against those of others. Any species co-operation, any species integration, is secondary to the particular organism's self-interests. And in any case, there are hardly likely to be species-wide secondary effects. Co-operation will, at most, be between relatives, or fellow population members. Generally selection pits organisms against each other (although not necessarily in a crude 'nature red in tooth and claw' fashion).[4]

Individual selection and the s-a-i thesis simply do not go together. What about obvious counters? Some biologists believe that group selection can work. This is true, but hardly makes the s-a-i thesis again compelling. Group selection supporters think it works for populations, not species, and no one denies the importance of individual selection. (See Wilson (1975a) and Wade (1978) for recent views on group selection.) Conversely, some biologists argue that the true 'individual' in individual selection is the gene, not the whole organism (Dawkins 1976). Does not my argument prove too much, suggesting that organisms should not be considered true individuals – which is clearly absurd? But, while this point does show that for some biologists the level of individuality does not necessarily stop at the whole organism level, no one denies that organisms (thanks to selection) are sufficiently well organized to be considered individuals in their own right. Richard Dawkins (1982), for instance, speaks of organisms as 'vehicles' which carry within them the units of selection, 'replicators' or genes. An organism, to such a biologist, is no less an individual than a BMW is to a racing driver.

Continuing with biological counters to the above critique, what about Steven Stanley's (1979) notion of species selection, where it is suggested that trends are a function of the success or failure of species? Again there is little help for the s-a-i thesis. Even if one accepts species selection, and many would not, the key operation of natural selection is with the individual. Drawing attention to the trends often seen in the fossil record, Stanley suggests that there is nevertheless a randomness about the members of new species with respect to a trend. Although a trend may (say) be from smaller to bigger, a new species in the line of descent could well have small-bodied members. (No doubt, if one persisted, one could devise some form of species selection where the group as a whole was significant. But its realization in nature is obviously another matter. See Arnold and Fristrup 1982.)

Perhaps the strongest biological case for the s-a-i thesis comes through the notion of a species as a number of organisms sharing a common 'gene pool', with shared types of genes being passed on from common ancestors (*i.e.* a kind of hybrid notion formed from several of the species concepts: Dobzhansky 1970, Dobzhansky *et al.* 1977). Here you might think we have the kind of integration required for individuality. Certain genes flow between the organisms of a species, and between no others. But this hardly acknowledges the key importance of individual selection. Moreover there are today strong questions about the biological importance, at the species level, of such genetic sharing. It was once thought that gene flow, between populations, was a key factor in keeping the organisms of a species alike. Now, it seems more likely that normalizing selection is the key causal factor. Species members sit on the top of the same 'adaptive peak': if they vary too much from the species norm, then selection wipes them out.

This downgrading of the significance of gene flow is a most important point, because (being itself one which comes from modern evolutionary theorizing) it strikes right at the heart of the claim that the s-a-i thesis (however counter-intuitive it may seem) must be accepted on the basis of modern biology. John Endler's (1977) already-classic study brings both theoretical and empirical evidence to bear demonstrating the restricted effects of gene flow. Basically, gene flow would be expected due to migration ('the relatively long-distance movements made by large numbers of individuals in approximately the same direction at approximately the same time', (Endler 1977, p. 182)) and dispersal ('the roughly random and nondirectional small-scale movements made by individuals rather than groups, continuously, rather than periodically, as a result of their daily activities'(1977, p. 181)). One would expect that migration would be a most effective way of uniting a species, even if widely dispersed; but as Endler points out this is rarely so, since migration is usually accompanied by return migration and organisms give birth in the place where they were themselves born. Such 'philopatry' has obvious adaptive virtues – birds may migrate to winter feeding grounds but they return to already established and proven breeding grounds.

Dispersal is a priori a less promising way in which gene flow might be greatly effective, and there are a number of reasons why its importance should not be overestimated: ethological, ecological, and physiological inadequacy of hybrids between distant species members; random loss of new-gene forms because they are rare; infrequency of long-distance travellers (as opposed to migrators); and more (1977, pp. 28-9). All in all, therefore, one should not overemphasise the unity of the species because of the supposed circulation of shared genes. (Similar points are made by Grant 1981a, 1981b, Levin 1981, and Mishler and Donoghue 1982. Although the point just made applies to both animals and plants, since the latter are spatially more fixed, expectedly the 'genetic integration' argument has always seemed less plausible to botanists.)

So, once again we come back to the individual organism and to its response to the environment (including fellow species members). If you take Darwinian selection seriously, you simply must reject the s-a-i thesis. Note that I am certainly not denying that the members of a species are frequently 'united' in having similar causal pressures, whether these be genetic or selective or something else. Of course they do. That is what makes them part of the same species. The question is whether this 'unification' is significantly more than similar causes. The s-a-i supporter has to say that it is, even to the point of the kind of integration we find in individual organisms, and this is what I deny. (Caplan (1980) rightly emphasized that same causes lead to same effects, and this accounts for species members being similar.)

WHY SPECIES ARE NOT INDIVIDUALS: CONCEPTUAL OBJECTIONS

Let us move on to the more conceptual type of objections to the s-a-i thesis. Crucial to the thesis is the claim that, logically,

109

a species can appear only once. If it dies, it cannot reappear. In this, species are just like paradigm individual organisms. As we have seen, Adolf Hitler cannot be resurrected and neither can the extinct species of pterodactyl.

I will leave Christians to fight their own battles about human bodily resurrection. As far as species are concerned, time and technology have shown the s-a-i claim wrong. Today, through recombinant DNA techniques and the like, biologists are rushing to make new life forms. Significantly, for commercial reasons the scientists and their sponsors are busily applying for patents protecting the new creations. Were the origins of organisms things which uniquely separate and distinguish them, such protections would hardly be necessary. Old life form and new life form would necessarily be distinct. Since apparently they are not, this suggests that origins do not have the status claimed by the s-a-i boosters. (See Wade 1979, 1980a,b.)

Also against the s-a-i thesis, in crucial respects it seems that it really does not treat species and their organisms very differently from the old way of treating species as classes, with members included according to the possession of certain required properties. Take an organism. How do you know that my hand is part of the individual, Michael Ruse? Because it is joined on – that's why! But my dog Spencer certainly is not joined on, in the same way, to the species *Canis familiaris*. So why do we want to say that he is part of the species? Because he descended from the original ancestors, along with the rest of the group – that's why!

Descent is starting to look very much like an essential property. Spencer is part of the group *Canis familiaris*. Indeed, we even seem to have a real/nominal distinction at work here. In Spencer's case, if challenged about his status, I can in fact produce papers attesting to parentage. But in the case of the other four-legged being that lives in my house, I have no such documentary evidence about origins. And yet I am as sure that Sesame is a cat, as that Spencer is a dog. Why? Because she looks and behaves like one. She miaows, purrs, keeps aloof, jumps from heights, stays up half the night, and is fastidiously clean. In short, she has all the identifying marks of cattiness (*Felis domestica*).

Clearly what I, and everyone else, am doing here is employing morphological and related criteria. Of course, Ghiselin and Hull recognize and appreciate the use of such criteria. They simply refuse to give such use any significant theoretical status (see Ghiselin 1981). What can this all mean, but that one is using nominal criteria, because real essence's descent relationships are unknown? Hence, for all the talk, the s-a-i thesis treats species as classes, with descent giving real essence and with morphology giving nominal definitions.

There are other objections you can raise against the s-a-i thesis. One is that it has controversial implications about the temporal limits of a species, and the possible evolutionary change that such a unit can encompass. So long as one gets no new group breaking off from an evolving lineage, one has one and only one species, whatever the change (just as one has one and only one organism, despite the change from caterpillar to butterfly). The cladistic school of taxonomy would accept this implication, but most biologists would not. They distinguish, for instance, between *Homo habilis*, *H. erectus*, and *H. sapiens*, despite the lack of branching. *Homo habilis* had a brain size of around 700 cc, much closer to that of a gorilla (*Gorilla gorilla*) at 500 cc, than to modern man at 1400 cc. Hence, species divisions are made – divisions which cladists and s-a-i supporters must ignore. (Cladism is discussed well in Eldredge and Cracraft 1980, and Wiley 1981.)

Indeed, the s-a-i thesis is more extreme than cladism. Cladists end a species as soon as there is any branching within the group. But if just a small population broke from a parent species, leaving the parent unaffected, one would have a situation very similar to an asexual organism budding off a small part. The parent remains. Similarly, the s-a-i thesis would have to count an analogous parent species the same original individual. Carried to the extreme, classification could become very difficult indeed.[5]

And, finally, let me point to an implication which has rather drastic consequences for the social sciences. Laws of nature generally do not refer to specific spatio-temporally bounded objects. They are rather timeless regularities in nature. But if species are individuals, then any claims about the organisms of species, restricted to the species, cannot be laws. In particular, any claims exclusively about human beings cannot possibly be laws. Hence, at one stroke, the social sciences, as they stand today, cease to be sciences in any worthwhile sense. To say the least, this is a somewhat drastic consequence.

The conclusion is clear. There is no absolute reason to treat species as individuals, and compelling reasons not to do so.

BUT WHAT THEN ARE SPECIES?

Either species must be groups, or they must be individuals. There is no third option. They are not individuals, hence they must be groups. But we have seen that species cannot have the absoluteness of Aristotelian natural kinds. Evolution makes this impossible. The question which therefore remains is whether we can raise species above the rather subjective level of Lockean kinds? Are species more than just artificial collections of organisms?

They are indeed. Moreover, the reasons that we rightly think that species are more than artificial collections, or why we think species are natural, are similar to reasons that we think there are natural groups encountered elsewhere in science, for instance in chemistry and geology.

The key reason that species are properly treated as natural kinds lies in that most distinctive fact noted earlier: the multiplicity of species concepts and of possible definitions of taxa names. To set the connection between naturalness and multiplicity, let us pull back for a moment and ask a general question about science. At what point is it in science that we feel we are on to something 'real'? When is it that we accept that we are not just dealing with figments of a creative scientist's imagination? The strong consensus is that the breakthrough comes when we put *together* two or more *different* areas of theory into one unified whole. If you have two different subjects, and they are joined, so that the one complements the other, then you are inclined to think that there's more than mere chance at work. Somehow, the unified theory tells you about the real. Such a coming together could not be mere coincidence, especially, if you can spell out the unification in terms of some overall theory (see Leplin 1984).

This unification, known philosophically as a 'consilience of inductions', was the one thing that Darwin always mentioned, when his theory was challenged. 'I must freely confess, the difficulties and objections are terrific; but I cannot believe that a false theory would explain, as it seems to me it does explain, so many classes of facts' (Darwin 1887, 1, p. 455). And, it remains important today. For instance, in the recent geological revolution, people accepted plate tectonics when they saw that different areas of geology were unified in the one theory (Ruse 1981e).

What about classification? Is there a possibility of some sort of consilience here, separating the natural or real from the merely arbitrary? William Whewell, who had the distinction of being both a professor of mineralogy and of moral philosophy, thought there was. A natural classification is one where different methods yield the same results. Particularly if the 'coincidence' turns out to have reasons behind it you feel sure that the classification cannot be just chance. 'The Maxim by which all Systems professing to be natural must be tested is this:– that the *arrangement obtained from one set of characters coincides with the arrangement obtained from another set*' (Whewell 1840, I, p. 521, his italics). (For more details on how Whewell used his ideas in mineralogy, see Ruse 1976b.) And modern philosophers agree with Whewell. Thus Hempel:

> The rational core of the distinction between natural and
> artificial classifications is suggested by the consideration
> that in so-called natural classifications the determining
> characteristics are associated, universally or in a high
> percentage of all cases, with other characteristics, of which
> they are logically independent. (Hempel 1952, p. 53. See
> also Schlesinger 1963, for more references to this criterion.)

Coming back to organic species, we see that we have a paradigm for a natural classification. There are different ways of breaking organisms into groups, and they *coincide*! The genetic species is the morphological species is the reproductively isolated species is the group with common ancestors. Moreover, there are reasons for the coincidence. As the zoologist Mayr points out, bringing several of the definitions together:

> The reproductive isolation of a biological species, the
> protection of its collective gene-pool against pollution by
> genes from other species, results in a discontinuity not

only of the genotype of the species, but also of its morphology and other aspects of the phenotype produced by this genotype. This is the fact on which taxonomic practice is based. (Mayr 1969, p. 28)

Note, moreover, that the coincidence between variously delimited species is not unexplained. Certain genes do lead to certain morphological effects. The consilience fits within overall biological thinking.

This consilience then is the reason why it is reasonable to think of species as natural kinds. Like the natural kinds of other sciences, they demand our attention, not because they represent some ultimate essentialist ontological carving up of the real world, but because they unite different criteria of division. They may not be Aristotelian kinds. But they are more than Lockean kinds (see also Ruse 1973c).

CONSEQUENCES

A number of questions arise. What about the real/nominal definition distinction? It vanishes – which is a good thing, because it is an outmoded Aristotelian holdover anyway. You might argue that genetic differences are more crucial than anything else. (Caplan 1980, 1981, and Kitts and Kitts 1979, argue just this.) But from an evolutionary perspective the genes do not have this kind of privileged status. If organisms do not have the right morphology, they will fail, no matter how superior their genotype. And in any case, a consilience is like a quarrel or a tango: you must have at least two parties. Hence it really does not make much biological or philosophical sense to say that genes are more essential than morphology, or that any other single feature is the 'true' essence of a species.

Do we still have laws about species members? I do not see why not. The solution I am offering affirms the existence of natural kinds, albeit not Aristotelian kinds. This means that it is still open to everyone to make universal claims about the members of particular species. And these claims can rise above the merely contingent or happenstance. This does not mean that every claim that has been made about human beings embodies genuine laws. I am not, for instance, defending the validity of every part of Freudian psychoanalytic theory. On the other hand, such theory is not being ruled out as a genuine science, on a priori grounds, before we even start.

Finally, let me make brief reference to some of the ongoing concerns that biologists have about species. I must emphasize that I am not trying to offer a quick and easy solution to every biological species query. Species which were difficult to evaluate before this essay will be as difficult to evalute afterwards. I am trying to show why biologists, generally speaking, think that species are natural. But the obverse side of the coin is that when difficulties arise and biologists no longer feel anything like as convinced of the naturalness of certain groups, the analysis offered above should show why. In particular, doubts about the reality of species should arise when the various ways of defining species come apart, and fail to coincide.

This is indeed the case. As mentioned earlier, botanists often find themselves less than convinced of the reality of many plant species. Why? Simply, because so often plant groups which morphologically and ecologically and in other ways seem to be good species, fail the test of reproductive isolation or some like thing. There is just not the consilience required for naturalness. Conversely, when there is isolation, there are sometimes few other differences. In such cases, and in analogous cases in the animal world, species – however drawn are not considered very natural. It is interesting to note how, in the case of so-called 'sibling species', where members of different reproductively isolated groups are morphologically similar, morphological differences are eagerly sought. Much relief is felt when such differences are found. (Grant 1981a, has an excellent discussion on the difficulties plants raise for species concepts. Mayr 1963, Dobzhansky 1970, and Dobzhansky *et al.* 1977, discuss in full detail difficulties arising in the animal world.)

Also, the analysis I have offered shows just why it is that biologists are far less convinced of the reality of taxa of higher levels than they are of species taxa. There simply is not the required consilience. Reproductive barriers are irrelevant, and there are no measures of morphological difference to coincide with genetic difference, and to coincide with evolutionary difference. If anything, the evidence is that such measures are impossible. Hence, although 'species are made by God, higher taxa are made by man'.

CONCLUSION

Ghiselin and Hull are surely right when they argue that we must break with Aristotelian essentialism in biology. After Darwin, such a position is otiose. But they go too far when they deny that biological species are natural kinds of any sort. They are such kinds, and the reason which makes us think any scientific claim goes beyond the merely hypothetical. It is that species are consilient.

NOTES

1 In my view, most of the modern supporters of natural kinds end up somewhere to the right of Aristotle (for example, Kripke 1972, Putnam 1975, Wiggins 1980). Frankly, I am not sure how far these modern thinkers really intend their ideas to apply to biology, since they generally do not bother to refer to the works of practising taxonomists, and at times show an almost proud ignorance of the organic world. Any comments I have to make against Aristotle apply equally against them. Dupré (1981) shows how ignorant most modern philosophical thinkers are about biological reality.

2 Other critics of the s-a-i thesis include Caplan (1980, 1981) and Kitts and Kitts (1979). Unfortunately, these critics revert to a modern-day, genetic, Aristotelian essentialism. I find myself agreeing with much in Hull's (1981) spirited response to them.

3 Could spatio-temporal contiguity alone count as the criterion of individuality? We surely think of the planet Earth as an individual on these grounds. But, while this may be true, we do not think of Earth as a biological individual, which notion is the focus of the s-a-i thesis. Incidentally, however, given plate-tectonic theory and the consequent claims about Earth's organization, a case might be made for Earth's geological individuality, transcending mere spatio-temporal contiguity.

4 There can be co-operation between organisms. But the point is that ultimately biology regards it as 'enlightened self-interest'. Hence at root we have tensions – separate reproductive strategies – between organisms. At times, for instance where mates are involved, these tensions break right out. See Trivers 1971, Wilson 1975a, Barash 1977, Clutten Brock 1982, and Ruse 1979b for more details.

5 An escape would be to embrace the neo-saltationary theory of 'punctuated equilibrium', supposing that one gets periods of stasis, followed by abrupt switches from one species to another.

See Eldredge and Gould 1972. The fears just expressed vanish. Wade (1980) and Mishler and Donoghue (1982) note just how tied up the s-a-i thesis is with this theory, and significantly the s-a-i thesis has been embraced enthusiastically by Eldredge (1985b). But there are serious queries about the position. See Gingerich 1976, 1977, Maynard Smith 1981, and Ruse 1982c for an overview.

Chapter Five

IS THE THEORY OF PUNCTUATED EQUILIBRIA A NEW PARADIGM?

Will a new synthesis emerge, signaling a true paradigm shift in the Kuhnian sense?(*Science* 1980)

The word 'paradigm' is just about the most over-used in the philosophical lexicon. In fact, professional philosophers tend to avoid it like the plague, and today it is much more commonly used by sociologists, scientists, and journalists such as Roger Lewin, quoted above. Part of the problem is that the word 'paradigm' is as slippery as the word 'God'. Everyone who uses it means something slightly different. Too frequently the term is used as a propaganda tool, bolstering the pretensions of some supposed major breakthrough: paradigm founder today, Nobel prize winner tomorrow, burial in Westminster Abbey the day after that.

All of this introduction is by way of explanation and (partial) apology. Several years ago when I first started thinking seriously about the paleontologists' theory of punctuated equilibria, I expected the inevitable. Some enthusiast would be hailing the theory as a new 'paradigm', suggesting that all who did not jump on the bandwagon were blind to the brilliance of the new science and doomed to an early extinction. The intellectual community was just waiting for the hold-outs to die, so it could go about its new proper business without guilt. Nor, to my morbid satisfaction, was I long disappointed. Sure enough, I found paradigm status being claimed (Stidd 1980). The strong implication was that all those who belittled the new position (as did an eminent evolutionist of my acquaintance who contemptuously remarked: 'Just hand-waving by the paleontologists,

Mike!') are simply irrational. All we were waiting for was the celebratory postage stamp.

Naturally enough, as a professional philosopher with strong Darwinian sympathies, I was happy to respond vigorously and (to my satisfaction) to make mincemeat of the claim (Ruse 1982c): in no way could one justify the conclusion that punctuated equilibria theory is a new paradigm; only one who is ignorant of philosophy and of science could be so naïve. Or so I said then. Now, I think I was wrong. At least, I think I was altogether too quick. It is not such a stupid thing to link together punctuated equilibria theory and new-paradigm status. It can, in fact, be positively illuminating, although whether the true link is quite what the enthusiasts have claimed is, perhaps, another matter.

In this paper, therefore, I intend to go back and answer the question properly. In the spirit of one who wants to learn new things rather than simply settle old scores, I am not going to make judgments about truth and falsity, 'better' and 'worse'.[1] I ask simply: 'Is the theory of punctuated equilibria a new paradigm?' And the answer depends, of course, on what you mean by 'punctuated equilibria theory' and what you mean by 'paradigm'. So let us start there.

WHAT IS THE THEORY OF PUNCTUATED EQUILIBRIA?

In recent years, philosophers of science who are interested in the history of science have been turning, with some enthusiasm, to evolutionary theory as a model for an adequate theory of change, including scientific change (Ruse 1986b). Organisms evolve. Theories evolve. Perhaps there are connections. As it happens, I myself am not overly excited by this particular strand of 'evolutionary epistemology'. I suspect there are some severely crippling disanalogies. Yet the idea is insightful in one way. As evolutionists know only too well, there is no such thing as your average or typical organism (Mayr 1963). Take any species, like *Homo sapiens*. There are big members, small members, black members, white members, bright members, stupid members. Is Stephen Jay Gould more typical than Brigitte Bardot? The question is meaningless.

Similarly, with theories, we tend to have the idea that there is such a thing as a theory, to which a number of people subscribe.

Supposedly there is a list of essential features (like the hypothetical list of essential features for species membership) and if you accept these features you accept the theory – but not otherwise. However, the evolutionary analogy highlights what anybody who tries to write on the history of science knows only too well. Theory nature – and theory allegiance in particular – is much more like the real nature of species membership. Different people believe very different things, and the same people believe different things at different times – and yet rally under the same banner.

I myself discovered this when I wrote a book on the Darwinian revolution and wanted to characterize 'Darwinian' (Ruse 1979a). Charles Darwin accepted natural selection and sexual selection, and applied his ideas to humans. Thomas Henry Huxley accepted evolution, applied it to humans, but was unenthused by selection. Alfred Russel Wallace accepted natural selection, had severe doubts about aspects of sexual selection, and pulled back from the evolution of humans. And so the story went. In the end, I had to be satisfied with some mushy sociological notion. A 'Darwinian' was someone who thought of himself as a Darwinian, or some such thing.

I feel a bit the same way about characterizing punctuated equilibria theory. Despite reassurances about uniformity and consistency – sometimes rather irritated reassurances – I see a number of different ideas and positions being staked out under the punctuationist flag. I say this in a sympathetic spirit, for one would expect a certain development of thought as people grapple to articulate a new idea. In a non-definitive way, let me suggest that there have been three basic phases to the punctuated equilibria theory, even in its short history of about a decade and a half.

The first phase was, fairly obviously, the form of the theory as it was announced at the beginning, most particularly in the paper by Niles Eldredge and Stephen Jay Gould, 'Punctuated equilibria: an alternative to phyletic gradualism' (1972). Without getting into details about whether this was truly the first articulation of the theory (Eldredge 1971 appeared just before, although it does not seem to me to have quite the panache of the jointly authored paper), we can say that this was the paper which got the attention and put the theory on the map. I think we can also say, with a reasonable amount of confidence, that this version of the theory was intended (at least, ostensibly) as a fairly straightforward extension of orthodox Darwinism (or, if you like, neo-Darwinism or the synthetic theory of evolution).

The authors noted that the fossil record does not show gradual change. Rather, it exhibits uniformity, with change across gaps or breaks. They noted also that modern evolutionary theory locates much, if not most, change at times of speciation, that such speciation is allopatric (involves geographical isolation), and that Mayr's founder principle is a crucial causal element. And they concluded that the empirical material, the fossil record, is explained by the theory's modern understanding of speciation, and that there is no need of subsidiary hypotheses about incomplete records and so forth. On the contrary, Eldredge and Gould professed themselves fairly happy with the fossil record. Furthermore, the authors underlined the conventionality of their position by stressing that their interpretation of natural selection was at one with that of everyone else. In particular, they wanted no truck with group or other deviant kinds of selection. Change in species comes down ultimately to change in organisms.

> The coherence of a species, therefore, is not maintained by interaction among its members (gene flow). It emerges, rather, as a historical consequence of the species' origin as a peripherally isolated population that acquired its own powerful homeostatic system. (We regard this idea as a serious challenge to the conventional view of species' reality that depends upon the organization of species as ecological uniters of *interacting* individuals in nature. If groups of nearly independent local populations are recognized as species only because they share a set of homeostatic mechanisms developed long ago in a peripheral isolate that was 'real' in our conventional sense of interaction, then some persistent anomalies are resolved. The arrangement of many asexual groups into good phenetic 'species', quite inexplicable if interaction is the basis for coherence, receives a comfortable explanation under notions of homeostasis. (Eldredge and Gould 1972, p. 223)

Remember, of course, that we are talking about the time when George Williams' *Adaptation and Natural Selection*(1966) was the most influential book of the day.

The second phase came as the decade drew towards its end, reaching its apotheosis in 1980 with Gould's notorious 'Is a new and general theory of evolution emerging?'(1980). There had to be something new going on here, for, far from portraying himself as an orthodox Darwinian, Gould assured us that the synthetic theory of evolution is effectively dead. Basically, I see a major de-emphasis of the importance of organic adaptation, with a consequent downplaying of the role of natural selection. Gould certainly was also starting to toy with the idea of macromutations of some sort (perhaps due to chromosomal rearrangements), with species' changes occurring in one or a couple of generations. Significantly, the father figure had changed from Charles Darwin to Richard Goldschmidt (1940), one of the few English-speaking saltationists of the past half-century, and intellectual rival to the synthetic theory's greatest living exponent, Ernst Mayr. (See Gould 1979b, 1980a, b; Gould and Eldredge 1977 – the last named a bit of a hybrid paper.)

I should say that Gould today (personal communication) categorically denies that he himself was ever a saltationist in Goldschmidt's or anyone else's sense. And it is certainly true that never in print did he enrol under a saltationist banner. What I think one can fairly say, however, is that Gould (especially) was starting to think of evolution's processes through a lens or filter of discontinuity (to use a metaphor). In his own mind, he was starting to highlight the essential abruptness of evolution, as opposed to its continuity – just as (say), a man falling in love might start to regard a woman in a new light, as a person of sexual attraction rather than as a lawyer or professor. Neither the continuity nor the profession is denied absolutely. They just no longer seem so important.

Then, fairly quickly after this, we move to the theory's third, and I think to date, final form. There is a pull-back from extremism, particularly with respect to the formation of new species. Thoughts of macromutation decline – indeed we are told that we were mean to think they were there in the first place – and we learn that although 50,000 years may seem like a long time to a fruit-fly geneticist, to a paleontologist it is but a blink of the eyelid (Gould 1982a, 1982b, 1983a, 1983b). In Gould 1982b, there is an explicit acceptance of macromutation, although it is disassociated from punctuated equilibria theory.

> Illigitimate forms of macromutation include the sudden origin of new species with all their multi farious adaptations intact *ab initio*, and origin by drastic and sudden reorganization of entire genomes. Legitimate forms include the saltatory origin of key features (around which subsequent adaptations may be molded) and marked phenotypic shifts caused by small genetic changes that affect rates of development in early ontogeny with cascading effects thereafter.(Gould 1982b, pp. 88-9)

However, although there may be pull-back there is certainly no retreat. Now we are presented with a *hierarchical* view of the evolutionary process. Down at the level of the individual organism we have natural selection working away, although much of its emphasis seems to be that of keeping things in line. (Below the level of the individual, at the level of the gene, we may have drift, working in ways that Japanese evolutionists suggest.) Adaptation counts, although it is only one thing, along with a lot of constraints on development and adult form. Change generally comes at time of speciation, and although full-scale mutations are out, there is a feeling that changes in rates of development could have significant and fairly instant effects. (There are interesting parallels here with E.O. Wilson's (1975a) much-criticized multiplier effect.) And what is important is that, as we pull back and look at species, we see that their evolution has patterns and dynamics of its own, at a level higher than that of the individual. These are patterns which sit upon and are connected to the individual but which are in no sense reducible to the individual. Against the well-known views of the geneticist Theodosius Dobzhansky (1951), macroevolution is *not* micro-evolution writ large.

A decade later, the prohibition against group selection is not as strong, and there are certainly suggestions that at the species level various effects might obtain that will make differences to success or failure. But, rather than group selection, more emphasis is put on the notion of 'species selection', where in some way species themselves act as units (this is very much in line with the philosophical claim that species are individuals and not classes) and where things like trends are seen as the working out of forces acting at this higher (*i.e.* above-the-organism) level.

I do not claim that these are three distinct theories or sub-theories. Steven Stanley (1979), for instance, has been promoting species selection since the 1970s, but I do claim that looking at punctuated equilibria theory in the way I have just sketched is not unfair to its history. What is unfair is the way in which from now on my discussion will be so heavily weighted towards the ideas of Stephen Jay Gould. Given what I have said about the supporters of theories, we have no right to think that others, including Gould's regular co-author Eldredge, share Gould's views. In fact, in Eldredge's latest books, *Time Frames* (1985a) and *Unfinished Synthesis* (1985b), I find much sympathy for what I have just characterized as the third phase of the theory. However, as Eldredge himself explicitly states (1985a, p. 161), there are differences between him and Gould, especially with respect to adaptation – with Eldredge being less critical of the notion. Also, Eldredge (1984) is obviously less than comfortable with any selection process that does not make central the individual organism. In view of what I shall have to say, these (especially the former) are no small differences. All I can do is plead that life is short, and confess to the real motive for my focus on Gould. I am a philosopher asking a philosopher's questions, and Gould more readily answers them than does anyone else. (Later, however, I shall return to the Gould/Eldredge differences.)

WHAT IS A PARADIGM?

How many names are there for God? I will brew down the meanings of 'paradigm' to four (Ruse 1981e). They are not accepted officially by anybody, including myself, but as with my analysis of punctuated equilibria theory they are offered in a non-critical, constructive spirit. I do not think my meanings are unfair to the spirit of the notion as popularized by Thomas Kuhn in his *The Structure of Scientific Revolutions* (1962).

First, we have what one might call the *sociological* sense of paradigm. This notion centres on a group of people who come together, feeling themselves as having a shared outlook (whether they do really, or not), and to an extent separating themselves off from other scientists. This sense of paradigm comes through in Kuhn's own characterization.

> Aristotle's *Physica*, Ptolemy's *Almagest*, Newton's *Principia* and *Opticks*, Franklin's *Electricity*, Lavoisier's *Chemistry*, and Lyell's *Geology* – these and many other works served for a time implicitly to define the legitimate problems and methods of a research field for succeeding generations or practitioners. They were able to do so because they shared two essential characteristics. Their achievement was sufficiently unprecedented to attract an enduring group of adherents away from competing modes of scientific activity. Simultaneously, it was sufficiently open-ended to leave all sorts of problems for the redefined group of practitioners to resolve. (Kuhn 1962, p. 10)

Second, we have the *psychological* level to a paradigm. People in a paradigm see things in a way differently from people not in the paradigm, especially from people in another paradigm. One group sees a duck, another sees a rabbit. One group sees dephogisticated air, the other oxygen. This is why, so often, in a change of paradigm – in a 'revolution' – you get something akin to a conversion experience. You do, literally, get a whole new way of looking at things.

Third, we have the *epistemological* level to a paradigm. Here, one's ways of doing science are bound up with the paradigm. What counts as a proper solution is defined by the paradigm, not to mention what counts as an interesting problem. For the pre-Copernican the difference between the inferior and superior planets was not a matter of great concern. For the post-Copernican, the difference was crucial and the fact that the heliocentric theory solves the problem whereas the geocentric theory does not, was taken as definitive positive evidence (Kuhn 1957). It is this way that epistemology functions in a paradigm-switch – asymmetrical from within to without the paradigm – that makes a revolution, in respects, beyond reason.

Fourth, we have the *ontological* aspect to a paradigm. At times, Kuhn suggests that the very world is, not so much defined by as created by a paradigm. What there *is* depends crucially on what paradigm you hold. For Priestley, there literally was no such thing as oxygen – it is not simply that he did not believe in such a thing as oxygen. In the case of Lavoisier, he not only believed in oxygen: oxygen existed. (Priestley, of course, did believe in dephlogisticated air: it was this that existed in his world.)

Others might want to put forward other notions of paradigm. These four will do for me, here.

IS THE THEORY OF PUNCTUATED EQUILIBRIA A NEW PARADIGM?

First, let me answer the easy parts to this question. As far as the sociological aspect is concerned, I see no reason to deny (more graciously, I am happy to allow) punctuated equilibria theory paradigm status. I doubt that either Gould or Eldredge would claim the same authority for 'Punctuated equilibria: an alternative to phyletic gradualism' as for Aristotle's *Physica*; but Kuhn grants – especially in his later writings (for instance, the second edition of *The Structure of Scientific Revolutions* (1970)) – that paradigms do not have to have earth-shattering significance. (Perhaps a slightly inappropriate metaphor in this context!) Certainly, the Eldredge and Gould work, and that of others, seem to have polarized evolutionists in such a way that punctuated equilibria theory has defining paradigm properties at the social level (see for instance Hallam 1977). Moreover, although I have no real direct evidence for this, my impression (from the speed with which people took defining positions) is that the sociological factors came into place fairly early on in the theory's history. (Sociology makes you think about the human factors. It is hardly a thesis of extreme methodological reductionism to point out that this aspect of paradigmhood relates fairly directly to the brilliant literary abilities of Eldredge and Gould – combined with the journalistic interests of the journal *Science*, in seeing all disputes and differences in stark black and white.)

Conversely, at the ontological level, I see little reason to push for paradigm status. I doubt if anyone seriously thinks that the theory has changed the fossil record, literally. Or, at least, let me put matters a little more moderately. If one accepts some sort of realist thesis, I doubt if one would be persuaded that the theory has changed the world. If one is a non-realist, ontological change to large extent collapses into questions of psychology and epistemology.

What then of the two middle categories of psychology and epistemology (which, from the paradigm point of view, are very closely related)? These are more difficult questions to answer. Towards such an answering, we must dig into the reasons why the theory of punctuated equilibria has attracted support. I think there are three plausible directions in which this enquiry might be pursued: empirical, political, and metaphysical (or theological, in a sense). I will take them in turn and then come back to the main question. As you will see, I believe the three directions in a sense correspond chronologically to the three phases of the theory, but I would not pretend that there are any tight connections.

THE EMPIRICAL ATTRACTIONS OF THE THEORY

Why would one accept the theory of punctuated equilibria? Is it because of the empirical evidence? Is it because of the fossil record? Somewhat paradoxically, the founders denied that this would be a reason: between punctuated equilibria theory and the older, 'phyletic gradualism', 'the data of paleontology cannot decide which picture is more adequate'(Eldredge and Gould 1972, p. 208). But I think we can take this with a pinch of salt. It is true that if you take seriously the Duhem-Quine thesis, about the relation between theory and evidence, you can always find some way of propping up a crumbling theory,[2] but, if one is prepared to stay with science, then there really does come a time when the facts start to bite. As a scientist, it is no longer reasonable to hang on to one position rather than another. Specifically with respect to punctuated equilibria theory, it is clear that Eldredge and Gould and fellow supporters really do think that the evidence matters. Perhaps you can go on arguing until you are blue in the face that the gaps in the fossil record are only artefacts of incomplete fossilization, but this becomes increasingly *ad hoc.* And how in any case do you explain stasis (lack of morphological change)? Are you seriously going to argue that all change occurs in the soft, non-fossilizable parts of organisms? Around us today, evolutionists see no reasons to draw distinctions between hard and soft. Why should they do so for the past?

The empirical evidence has certainly been intended as important support for the punctuated equilibria position. I should add incidentally that by 1977 (in their follow-up paper, 'Punctuated equilibria: the tempo and mode of evolution reconsidered') Gould and Eldredge themselves were busily denying that they had ever belittled the importance of the empirical evidence. Supposedly, all they had suggested is that the facts tend to be 'theory laden', that is, we tend to look at their world from one viewpoint or another.[3]

The evidence has been intended as support, and moreover it would be churlish to deny that it has been effective support. One thinks particularly of Peter Williamson's (1981) pertinent findings about the evolution of molluscs in the lakes of East Africa. Everyone seems to agree that we do here have the sorts of goings on that manifest a punctuated picture of evolutionary change. On the negative side, Gould's pulling back from the extreme discontinuities of the second phase of the theory was in major part a function of the evidence. On the one hand, he was under fire from other evolutionists, particularly from the geneticists, who argued flatly that there is just not the kind of empirical base that he required. On the other hand, those same geneticists gave him an empirical escape route – or, if you like, a more fruitful path of discovery. They pointed out how much change there can be in 50,000 years, even with slow-breeding organisms (Stebbins and Ayala 1981).

So we can certainly give the empirical evidence a place in the battle over punctuated equilibria theory. However, this said, it has been clear from fairly early on that – whatever the founders may really have intended – the empirical evidence, particularly in the fossil record, is truly not the ultimate court of appeal. Apart from all the questions about whether our knowledge of the fossil record tells more of the fossils or of the classificatory techniques of paleontologists, there are two main problems. First, the paleontologists disagree about how often the fossil record exhibits stasis and how often it exhibits gradualism. Just about everyone allows that we get a bit of stasis. Just about everyone allows that we get a bit of gradualism. But, how precisely one understands a 'bit' is quite another matter. All one can say is that there may be some final answer to this debate. Perhaps at some future convention paleontologists will agree that some

85.021% of cases show stasis but, at this time, the final answer has not been given by competent paleontologists. (Gould and Eldredge 1977 shows well the difficulties with the evidence. See also Hallam 1977, and the recent Gould and Eldredge 1986.)

The second problem is that a lot of the supposedly crucial cases turn out to be highly ambiguous. Take, simply, the case of *Homo sapiens*. Did our evolution exhibit punctuated equilibria? Some say 'yes', some say 'no'. As a detailed discussion by the late Glynn Isaac (1983) shows only too well, there really is no definitive answer.

I conclude, therefore, that the empirical evidence is far from definitive, and I back this up by an (admittedly undocumented) sense that, in recent years, opponents have not really been striving all-out to find crucial test cases. Such cases have tended to become a bit like prayer: comforting to believers. Do not misunderstand me. I am not saying evidence could (in principle) never be effective. Moreover, empirical knowledge was crucial in the formulation of punctuated equilibria theory. Claims about allopatric speciation are derived from our knowledge of the world. What I am saying is that we need to keep looking for the theory's hold on good biologists.

TWO POLITICAL ATTRACTIONS OF THE THEORY

You can mean a number of things by 'political'. At one level, you can mean the politics internal to science, or even internal to biological science. At another level, you can mean the politics of the big world, outside. Let us take them in turn.

With respect to internal politics, I am sure that they have played (and do still play) a factor. I do not know how much of a factor, and I expect that the answer varies from person to person. I strongly suspect that paleontologists suffer from inferiority complexes, recognizing and resenting the fact that other evolutionists, especially geneticists, make the running, and resenting even more that other evolutionists think this a rightful ordering of things. ('Hand-waving by the paleontologists, Mike!') Supposedly, it does not really matter what happens in the fossil record: it has got to fit in with the chromosomes. And not vice-versa. When Gould toys with the notion of macromutations, he is slapped down, whereas no paleontologist would dare challenge (say) William Hamilton (1964a, 1964b) when he proposes kin selection.

I suspect that, at the beginning, there was simple pleasure at making an important point and showing usefulness. Remember, the first phase of punctuated equilibria theory was orthodox neo-Darwinism and the claim to fame was that paleontology could do really useful work in confirming the theory. The cry seemed to be: 'Don't think of us as a weak sister. We can play a full role in supporting the cause.' Now, as the theory has matured to the third phase, the claim is much stronger: 'There are aspects – vitally important aspects – of the evolutionary process that you just cannot understand without the aid of paleontology. You would not even recognize them without paleontology. Hence you ignore us at your peril.' This, it seems to me, is a powerful political reason for being a punctuated equilibria supporter. (If you think I exaggerate these matters, then look at Gould's 'Irrelevance, submission and partnership: the changing role of paleontology in Darwin's three centennials and a modest proposal for macroevolution' (1983a). The main title tells all. Significantly, this paper was given at the Darwin Centennial Conference in Cambridge, England, 1982, when Gould – and paleontology – were on display in front of the rest of the evolutionary world.)

Turn next to the politics of the world external to science. Until recently, most would have pooh-poohed the idea that (general) politics could affect science, making some theories more attractive than others. However, thanks to the work of historians (like Gould himself) we now know that this is naïve thinking – perhaps such denial is itself derived from a political stance which necessitates the objectivity of science (particularly your own science). Certainly, Gould (1981) has been one of the forerunners in recent years in arguing that aspects of contemporary evolutionary thought are impregnated with a certain political ideology. I refer to the claim that human sociobiology is capitalist libertarianism, by another name (Allen *et al.* 1977).[4]

What about punctuated equilibria theory? What cannot be denied is that Gould has admitted, if not to being a Marxist, to having learnt his Marxism at his Daddy's knee (Gould and Eldredge 1977), that he has thought it pertinent to bring this fact into his scientific work (on the nature of paleontological theories), that he has admitted to seeing the world as functioning according to the laws of dialectical materialism, and that he has praised punctuated equilibria theory precisely because it fits with such a dialectical world-picture (Gould 1979b).

When he was questioned on these links (in print in 1981 and verbally in 1982 by the paleontological gadfly Beverley Halstead, at the Cambridge Darwin Symposium) Gould tended to downplay their significance, but they were there – especially around 1980, just when Gould was promoting the second phase of the theory. This, of course, was the most discontinuous version of the theory and with reason might be thought that most in tune with a Marxist world-picture. Remembering also that Gould was a leader in the fight against human sociobiology – which fight was carried on most strenuously by left-wing groups like the Cambridge (Mass.) Science for the People collective, of which Gould was a member (Ruse 1979b, Segerstralle 1986) – it is worth recalling that the fight against this was most intense towards the end of the last decade.

And it is interesting to note how the sociobiological controversy and the punctuated equilibria controversy were drawn together. If the punctuated equilibria theory be true of humans, then essentially all of the change leading to today's forms happened in one leap right at the beginning of our species' origin. Since then, there has been little or no change, especially not change of a significant adaptive nature. Obviously, all of this blows holes in any kind of extreme sociobiological thesis, which, through the multiplier effect, sees major adaptive changes occurring virtually overnight, within the same species. The links are there.

I think it is worth mentioning also that 1981 was the year Gould published *The Mismeasure of Man*, a passionate treatise on the history of hereditarian claims about intelligence, a significant theme of which is the way in which such views have been used to discriminate against Jews, especially in America in the first part of this century. I speak of 'passion' in part because, of all Gould's books, this is the one which got widely criticized, particularly by historians of psychology. This perhaps suggests that strong motives framed its nature. (I should go on record as saying that I wrote a highly favourable discussion review, Ruse 1982a.)

It does not seem to me entirely implausible to suggest that Gould's passion against human sociobiology was linked to the fear that it was yet another tool which could be used for anti-semitic purposes. I did ask Gould about this in December 1981, when we were both witnesses for the American Civil Liberties Union against the Creationists in Arkansas. He himself did not entirely repudiate the idea, but inclined to think that the opposition stemmed more from Marxism, and as it so contingently happens, many American Marxists are from Eastern European Jewish families. Perhaps both factors were involved.

Again, do not misunderstand what I am saying. I am *not* saying that punctuated equilibria theory is part of a communist plot. Most of its supporters (like Eldredge) are not Marxists. Still less am I saying that critics of the theory are guilty of anti-semitism. I have already pointed out how dangerous it is to believe automatically that one person's theory is another person's theory, even if they fall in the same camp. And I would argue this point even more strongly for motives. All I am saying is that there is a complex set of threads which bind people's motives to their support for scientific theories, and some of those motives I have just been detailing may have lain – may still lie – behind the support that some people have for punctuated equilibria theory.

As with the empirical evidence I do not argue that this is all of the story. I am inclined to think that, at most, we are still dealing with secondary causes. The direct connections between support for punctuated equilibria theory and opposition to human sociobiology are frail. But they are there. Moreover, as I shall explain in a moment, I think we are starting to hint at some of the really deep reasons for the support of punctuated equilibria theory.[5]

THE METAPHYSICAL ATTRACTIONS OF THE THEORY

Let me start this section with a question, which I feel confident of answering (Ruse 1979a). What was Darwin's big achievement, or more precisely, what was it that Darwin thought of as his big achievement? Most obviously, it was establishing the fact of evolution – and no one (except the Creationists) would deny him credit for this. But in Darwin's own mind, at least, there was more. After all, by 1859 (the year of the publication of the *Origin*) many others had proposed the idea of evolution. For Darwin, his big achievement was his mechanism of natural selection. And in order to know why Darwin thought it a big achievement, we have to know (as always with answers) what the question was. In Darwin's case, it is made clear with his first introduction of natural selection.

> Before entering on the subject of this chapter, I must make a few preliminary remarks, to show how the struggle for existence bears on Natural Selection. It has been seen in the last chapter that amongst organic beings in a state of nature there is some individual variability; indeed I am not aware that this has ever been disputed .... How have all those exquisite adaptations of one part of the organisation to another part, and to the conditions of life, and of one distinct organic being to another being, been perfected? We see these beautiful co-adaptations most plainly in the woodpecker and misseltoe; and only a little less plainly in the humblest parasite which clings to the hairs of a quadruped or feathers of a bird; in the structure of the beetle which dives through the water; in the plumed seed which is wafted by the gentlest breeze; in short, we see beautiful adaptations everywhere and in every part of the organic world.

> All these results, as we shall more fully see in the next chapter, follow inevitably from the struggle for life. Owing to this struggle for life, any variation, however slight and from whatever cause proceeding, if it be in any degree profitable to an individual of any species, in its infinitely complex relations to other organic beings and to external nature, will tend to the preservation of that individual, and will generally be inherited by its offspring. The offspring, also, will thus have a better chance of surviving, for, of the many individuals of any species which are periodically born, but a small number can survive. I have called this principle, by which each slight variation, if useful, is preserved, by the term of Natural Selection. (Darwin 1859, pp. 60-1)

The point was that natural selection was to speak to *adaptation* – the hand and the eye. This was the question that Darwin wanted answered. Moreover, from all sorts of evidence, we know precisely why it was that Darwin thought adaptation so significant a feature of the organic world. He was immersed in the natural theology of early nineteenth-century Britain. Although he was to reject the Great Designer in the sky of Archdeacon William Paley (1802), author of the authoritative texts on the subject, Darwin accepted completely the premise of the natural theologians that the organic world was *as if* designed. (Incidentally, Darwin was a believer for most of his life, including the time of writing the *Origin*.)

This is the key point to understanding Darwin, even if you yourself doubt the efficacy of natural selection. Furthermore, it is the key point to understanding the thread of continuity from Darwin to the present. Consider, for example, the work of Sir Ronald Fisher, especially his *The Genetical Theory of Natural Selection* (1930); or Sir Julian Huxley *Evolution: The Modern Synthesis* (1942) or Theodosius Dobzhansky *Genetics and the Origin of Species* (1951) (especially after what Gould has referred to as the 'hardening' of the synthesis); or, in our day, G.C. Williams' *Adaptation and Natural Selection* (1966), or the whole socio-biological movement. Everybody thinks in terms of design – this was what Gould meant by 'hardening'. Williams puts the point explicitly:

> Whenever I believe that an effect is produced as the function of an adaptation perfected by natural selection to serve that function, I will use terms appropriate to human artifice and conscious design. The designation of something as the *means* or *mechanism* for a certain *goal* or *function* or *purpose* will imply that the machinery involved was fashioned by selection for the goal attributed to it. When I do not believe that such a relationship exists I will avoid such terms and use words appropriate to fortuitous relationships such as *cause* and *effect*. (Williams 1966, p. 9)

Moreover, although no one (including Darwin – or Paley for that matter) would argue that all organic features are useful, the presumption is that when we are looking at features, especially complex ones, adaptation is there unless proven otherwise (which otherwise might well involve adaptation at one step removed, like past function). Richard Dawkins says it all:

I agree with Maynard Smith (1969) that 'The main task of any theory of evolution is to explain adaptive complexity, i.e. to explain the same set of acts which Paley used as evidence of a Creator'. I suppose people like me might be labelled neo-Paleyists, or perhaps 'transformed Paleyists'. We concur with Paley that adaptive complexity demands a very special kind of explanation: either a Designer as Paley taught, or something such as natural selection that does the job of a designer. Indeed, adaptive complexity is probably the best diagnostic of the presence of life itself. (Dawkins 1983, p. 404)

Now, relate this to punctuated equilibria theory. No one denies that natural selection plays a role. In the early versions, especially, selection plays an important role. No one denies adaptation. But if punctuated equilibria theory does anything, it downplays the significance of natural selection and (especially) adaptation. Functions really do get squeezed down, as change is compressed into short periods. In addition – at least in some versions of species selection – when a new species is produced there is an element of randomness about adaptive value.

I do not claim that a new force of evolutionary change has been discovered. Selection may supply an immediate direction, but if highly constraining channels are built of nonadaptations, and if evolutionary versatility resides primarily in the nature and extent of non-adaptive pools, then 'internal' factors of organic design are an equal partner with selection. (Gould 1982c, p. 384)

This downplaying of the significance of adaptation is not a contingent by-product of the theory of punctuated equilibria. Gould, especially, has been a persistent critic of adaptationism for fifteen years. He accuses adaptationists of producing 'just so' stories, like those of Rudyard Kipling (Gould 1982a). He faults adaptationists for 'Panglossian optimism', because they see design-effects in all sorts of non-existent places (Gould and Lewontin 1979). He cites all sorts of examples of supposed adaptations which are not really adaptations at all. One neat example is the much-enlarged clitoris of the female hyena, which might lead you to think it functions as a quasi-penis. Gould argues it is more plausibly a side product of testosterone levels (Gould 1983c). And repeatedly, Gould has suggested that change might be merely contingent on change of rates of development, and thus without any immediate purpose (Gould 1977a).

Is this all just negativism on Gould's part – or, at most, that he saw that his punctuated equilibria theory downplays adaptation and so he had made a virtue out of necessity? There could well be a negative element here, especially in so far as the attack on adaptationism represents an attack on the highly adaptationist human sociobiology. This connection has been made, quite explicitly (Gould 1981). However, it would be a bad mistake simply to think in negative terms: that punctuated equilibria theory (particularly in its final phase) defines itself negatively by belittling a key plank in the Darwinian programme. Rather, we would do better to start gathering up the strands of evidence we have had thus far, particularly the European connection of which Marxism is a major intellectual outpouring, add to it Gould's expressed enthusiasm for the Germanic trans-cendentalist notion of *Baupläne* (blueprints for bodily structure) as well as for morphologists like D'Arcy Thompson who were likewise keen on underlying structures (Gould 1971), and finish off with a strong dash of Gould's reading of history, particularly in his *Ontogeny and Phylogeny* (1977a), where he expresses strong admiration for various lines of European thought.

The answer which then emerges is that we are faced with a thinker from a biological tradition different from the utilitarian adaptationism (to use the technical term) of the Darwinian. We are looking at somebody, and, in punctuated equilibria theory, at somebody's theory, whose roots go back to the transcendental idealism of Goethe and other early nineteenth century thinkers. In the words of that great historian of biology, E.S. Russell (1916), we are looking at an emphasis on *form* rather than on *function.* We are looking at people whose emphasis is on structure, shared between animals, because of underlying laws of constraint.

This is not to be saying something negative. In suggesting that punctuated equilibria theory stands in the tradition of Goethe, I am not trying subtly to suggest that Gould is not a genuine evolutionist. We are talking now about mechanisms, and in any case, I suggest also that Darwin stands in the tradition of Paley, a lot less of an evolutionist than Goethe ever was. In fact, there are the two traditions which go right across the evolutionary divide. (See Russell 1916, Mayr 1982, and Appel 1987 for full details.) On the functionalist side, we have Paley (who goes back to the external teleology of Plato) and the French father of comparative anatomy, Georges Cuvier (who goes back to the internal teleology of Kant and Aristotle), and then we move up to the Darwinians. On the formalist side, we have the transcendentalists, not just Germans like Goethe and Oken and von Baer (I would include him here) but also the French like Geoffroy St Hilaire and the English, like Richard Owen, the notorious evolutionist Robert Chambers, T.H. Huxley in respects (his attacks on transcendentalism were more than anything personal attacks against Owen, than anything else), and the Swiss-American Louis Agassiz. To this side, now, I would add supporters of punctual equilibria theory – at least, supporters of the later versions.

It must be noted that virtually no one belongs to one side exclusively. Darwin was always extremely proud of the fact that he could explain *Baupläne*, as they manifest themselves in repeated patterns (homologies) from organism to organism (Ruse 1979a). This 'unity of type', he thought, fell out from his overall evolutionism. The point is that for Darwin these *Baupläne* are frames on which adaptations can be hung. Conversely, someone like (say) Owen (1849) never denied adaptation. No more does Gould. It is all a matter of emphasis and relative importance.

It should also be noted that among evolutionists these two sides have traditionally been associated with different attitudes to the tempo and mode of evolution (to use G.G. Simpson's (1944) happy phrase). Darwinism can certainly accept, indeed expects, different rates of evolution, from very slow or non-existent all the way over to very rapid, the kind that might not get recorded in the fossil record. However, for Darwin and his fellow thinkers, the essence of change has to be gradual because otherwise organisms would be liable to get out of adaptive focus. This is the real reason why Darwin repudiated any kind of saltationism. (This can be seen when looking through the early notebooks and first drafts of the theory. Darwin toyed with saltations in the early days, but then realized that they pose great difficulties for adaptation. See Ruse 1979a, Ospovat 1981.)

Conversely, although the transcendentalist tradition has never excluded gradualism entirely, such gradualism has never been seen as essential and there has always been a place for 'jumps', however these might be interpreted (Reif 1983). Chambers (1844), for instance, was an out-and-out saltationist. And T.H. Huxley (1860) chided Darwin for tying himself too tightly to the motto: '*natura non facit saltum*'. Punctuated equilibria theory is taking a stand well within its tradition. (This is, of course, also the tradition of the German philosophers like Marx and Engels. This is why I prefer to see the various parts of Gould's thinking as connected through a general metaphysical position, rather than tacked together piecemeal as the particular controversy demands.)

Additional support for my claim that punctuated equilibria theory's real roots lie in the transcendentalist tradition, and that the reason it (especially in its final form) appeals to certain biologists is that they too turn or are attracted to that tradition, comes from the attitude taken towards human sociobiology. Let me note that the simultaneous opposition to human sociobiology makes much sense because (apart from the direct links) this opposition is seen as part and parcel of a general attitude, given that human sociobiology is so much an extension of neo-Darwinism (Ruse 1987a). And finally, let me quote the conclusion to one of Gould's essays (published in 1983). He is talking about adaptation:

> I have said nothing about German biology because it has generally held a view of adaptation outside the scope of this essay (Rensch and other synthesists notwithstanding). Adaptation is seen as real but superficial, a kind of jiggling and minor adjustment within a *Bauplan* evolved by some other mechanism – not, in any case, a general mechanism (by extrapolation) for evolutionary change at higher levels....

> The adaptationist tradition, on the other hand, has been an English pastime for at least two centuries. If continental thinkers glorified God in nature by inferring the character of his thought from the laws of form linking his created species, or incarnated ideas (as Agassiz maintained), then Englishmen searched for him in the intricate adaptation of form and function to environment – the tradition of natural

theology and Paley's watchmaker. Darwin approached evolution in a quintessentially English context – by assuming that adaptation represented the main problem to be solved and by turning the traditional solution on its head. Few continental thinkers could have accepted such a perspective, since adaptation, in their view, was prevalent but superficial. The centrality of adaptation among English-speaking evolutionists in our own times, and the hardening of the synthesis itself, owes much to this continuity in national style that transcends the simple introduction of evolutionary theory itself. One might say that adaptation is nature's truth, and that we had to overthrow the ancient laws-of-form tradition to see it. But one might also say that twentieth-century panselectionism is more a modern incarnation of an old tradition than a proven way of nature. (Gould 1983b, pp. 90-1)

One might also say that twentieth-century punctuated equilibria theory is more a modern incarnation of an old tradition than a proven way of nature.[6]

IS THE THEORY OF PUNCTUATED EQUILIBRIA A NEW PARADIGM?

I ask the question again, thinking now about the second and third senses of paradigm, the psychological and the epistemological. In their original paper (1972), Eldredge and Gould were a little bit coy. Explicitly (in a footnote) they denied that they were in the paradigm-invention business. Yet with all their claims about the facts not deciding the issue, one rather senses that they would not have been broken-hearted had someone brushed aside their modesty and contradicted them. However, at this point in the theory's history, I really do not think it would be appropriate to talk of (new) 'paradigm', especially not in the epistemological sense. The whole point about punctuated equilibria theory when it was first introduced was that it was an extension, or rather, correct application of an *already-existing* paradigm, namely orthodox neo-Darwinism. It was certainly agreed that one may now see things differently. But this claim was not based on a supposed paradigm switch. Rather, the claim was that only now was one using properly the paradigm in hand.

However, as the theory has matured, particularly through to its third phase, I would argue that the case for paradigm status (in the second and third senses) becomes much stronger. This is especially true if you grant what I have been trying to show in the last three sections, namely that the ultimate appeal of the theory lies in its transcendentalist roots. In a way, there are two metaphors which colour one's picture of the organic world. On the Paley/Darwinian side, there is the world as an object of design, but not just any kind of design – *artefactual* design. Organisms are seen as like artefacts. The eye is a telescope. The heart is a pump. The fins on the back of stegosaurus are heat-regulating devices akin to those found in hydroelectric stations. On the transcendentalist side, you can still have the world as an object of design, but not just any kind of design – *crystalline* design. Organisms are seen as like crystals. There are various laws which determine possible patterns and these are repeated throughout various branches of organism.[7]

All of this, it does seem to me, points to something (at both psychological and epistemological levels) much like a Kuhnian paradigm difference. In looking at organisms, or their fossil remains, one is looking at different things (not necessarily in some ultimate ontological sense). And certainly, this leads to different methodological judgements and directives. The orthodox Darwinian searches for those gradual linking sequences in the fossil record, and tries to understand change in terms of adaptive response to selective pressures. The punctuated equilibria supporter knows that these links will be rare or non-existent, and their search a waste of time, and he or she looks for other reasons behind change (or non-change). This supporter also thinks it worthwhile to seek answers to questions about higher levels of the evolutionary process, questions which I am not sure the orthodox Darwinian recognizes as questions at all. (Look at Maynard Smith 1983.)[8]

Grant then that there is indeed something going on here which looks like a paradigm (or paradigm difference) in action. People (like my former self) who dismissed the idea were wrong, and missing something rather interesting to boot. I want to conclude by making three qualifying, or clarifying, points. First, note that what we have are things which are rather softer than Kuhnian paradigms. For Kuhn, it is one or the other, but not both. Now it is a duck. Now it is a rabbit. There are no duck-rabbits. The artefact/crystal metaphors are a little different in that, as we have seen, there is room for overlap, with everybody at least appreciating the other side, somewhat. There is not quite the stark contrast that Kuhn supposes.

Second, we see how the two positions have had a long parallel history. Now one position predominates, now the other; here one position excites biologists, here the other. But for two hundred years both have been around. We do not get the kind of sequential story that *The Structure of Scientific Revolutions* leads us to suspect. Indeed, I am not sure just how much support our story gives to the very notion of a 'revolution' in science. I am not denying that revolutions occur, but they are certainly not as Kuhn portrays them. The coming of punctuated equilibria theory was no revolution, in the sense of a new paradigm. But, before you think I am simply belittling the work of Eldredge and Gould and others, let me point out that the coming of the *Origin of Species* was no revolution, in this sense, either! I should add that I do not think this is a point that should be accepted in glum negative silence, but as a spur to thinking again about the nature of scientific change.[9]

The third point follows on the other two. What of the future? May we expect a resolution? Will one paradigm or metaphor ultimately prove triumphant? I confess that I am not quite sure of the correct answer here, although I would remind you that no one is looking for total victory anyway. Against Kuhn, I would not rule out a priori the possibility of the empirical evidence being fairly definitive, even though it has not been so yet. Suppose some new technique of studying the fossil record yielded masses of new pertinent evidence. This might close the case. Although I have to admit that on the other hand it might not. Gel electrophoretic techniques have given us much new information about genetic variation, but the disputes go on (Lewontin 1974).

Another possibility is that people will bury the hatchet and start to balance their evolutionary thinking between the two traditions. Gould, at times, seems to yearn for such a 'pluralistic' account of change. (See, for instance, Gould 1983b.) Given human nature, however, I wonder if this is not wishful thinking.

Perhaps the two traditions will simply continue on into the future, with each side having supporters. In the realm of paleontology, there will be strict adaptationists and strict punctuationists (or whatever the views are then called). The consolation is that this is not necessarily a bad thing. Indeed, the creative tension that the traditions cause can have a positively stimulating effect. The history of paleontology in the past fifteen years and the way that the theory of punctuated equilibria has brought the field alive is, surely, proof enough of this fact.

EPILOGUE

Several years ago, I spent a year in the Museum of Comparative Zoology at Harvard. On my first day there, I was being shown the ant collection by that already-mentioned distinguished evolutionist who was so contemptuous of punctuated equilibria theory. He held up some particular specimen of ant and started to hold forth on some intricate adaptation. 'What a remarkable example of design!' he cried out. Then he stopped, looked at me sheepishly and grinned. 'For goodness sake! Don't ever tell Steve Gould that I said that.'

I am grateful to Niles Eldredge and Stephen Jay Gould for comments on the first version of this paper. Although both had critical things to say – Gould particularly was keen to stress that he had never thought of himself as a saltationist – I think I can fairly say they were not unsympathetic to my analysis.

NOTES

1 Two philosophers who have discussed the theory are Sober (1984) and Thompson (1983).
2 The Duhem-Quine thesis, or D-thesis, points out that if you have a hypothesis H, supposedly leading to an empirical observation O, but not-O obtains, then instead of denying H (which is what *modus tollens* and Popper's principle of falsifiability demand) you can always deny an auxilliary hypothesis A, for in science you never get observations from one hypothesis alone but always in conjunction with many others (for instance, about the realiability of one's equipment).
3 This is not the only occasion in the short history of punctuated equilibria theory that minds are changed although it is

pretended that no shift has occurred. I suppose that when you are fighting for your views, any shift is taken as weakness. But I wish people would see that changing or modifying one's thinking can be a virtue as well as a fault.

4 This was the work of a collective, which included Gould.

5 Since Gould has strongly denied significant links between his scientific position and his political beliefs, it seems fairest to give one of the more explicit discussions where the two are linked and then a repudiation. First, from Gould and Eldredge (1977):

When Darwin cleaved so strongly to gradualism – ignoring Huxley's advice that he did not need it to support the theory of natural selection – he translated Victorian society into biology where it need not reside ... We mention this not to discredit Darwin in any way, but merely to point out that even the greatest scientific achievements are rooted in their cultural contexts – and to argue that gradualism was part of the cultural context, not of nature.

Alternate conceptions of change have respectable pedigrees in philosophy. Hegel's dialectical laws, translated into a materialist context, have become the official 'state philosophy' of many socialist nations. These laws of change are explicitly punctuational, as befits a theory of revolutionary transformation in human society. One law, particularly emphasized by Engels, holds that a new quality emerges in a leap as the slow accumulation of quantitative changes, long resisted by a stable system, finally forces it rapidly from one state to another (law of the transformation of quantity into quality)....

In the light of this official philosophy, it is not at all surprising that a punctuational view of speciation, much like our own, but devoid (so far as we can tell) of a reference to synthetic evolutionary theory and the allopatric model, has long been favored by many Russian paleontologists ... It may also not be irrelevant to our personal preferences that one of us learned his Marxism, literally at his daddy's knee.

Then, in a letter to *Nature* (26 February 1981), responding to Beverley Halstead's charge which Gould characterized as: 'simply silly beyond words, that punctuated equilibrium, because it advocates rapid changes in evolution, is a Marxist plot,' Gould wrote:

For Halstead's second charge, I did not develop the theory of punctuated equilibrium as part of a sinister plot to foment world revolution, but rather as an attempt to resolve the oldest empirical dilemma impeding an integration of palaeontology into modern evolutionary thought: the phenomena of stasis within successful fossil species, and abrupt replacement by descendants. I did briefly discuss the congeniality of punctuation change and Marxist thought (*Paleobiology*, 1977: 145) but only to illustrate that all science, as historians know so well and scientists hate to admit, is socially embedded. I couldn't very well charge that gradualists reflected the politics of their time and then claim that I had discovered unsullied truth. But surely Halstead, who has done some statistics in his day, knows that correlation is not cause. If I may make a serious point: I grew up frightened in a leftist household during the worst days of McCarthyism in America; and I know that what seems peripheral or cranky today can become a weapon tomorrow (consider the current creationist surge in America). May we avoid red-baiting; it may not always be harmless. (289:742)

There was no 'plot' but there was a bit more than 'correlation'. Why the retreat? In part, I suspect, because punctuated equilibria theory by the early 1980s had a high profile, and did not need the 'taint' of controversial ideology. In part, it was because the fight against human sociobiology likewise did not want to supply the opposition with such a ready weapon, and in part, because the fight against creationism was not helped by the connection. I remember in Arkansas that, thanks to my having drawn the connection, both Gould and I were cross-examined on this by the assistant attorney general (appearing for the state in defence of the creationist law). In part, because (as Gould says above) charges based on political ideology have such painful memories.

6 Let me return again to the worry that I am dealing less with punctuated equilibria theory *per se* than with one man's (Gould's) version of it. Could it not be argued that although I may put my finger on why Gould believes in the theory, this is not necessarily the reason why others do. Indeed, if I may invoke the old distinction between the context of discovery and the context of justification, you might argue that just about everything I am saying is irrelevant to the reasons why the theory is accepted. In response, all I can say is that this line of argument virtually gives the case away, by definition. Only empirical evidence is allowed to count, and that is the end of matters. I would myself challenge the discovery/justification dichotomy, and as part of my case I would note that Gould's position is stronger against adaptationism than is Eldredge's, and since they both have access to the same evidence would claim that the difference has to be explained in Gould's deeper roots in continental philosophy and science. (I am interested to note, incidentally, that although *Time Frames* contains several references to Gould's early work, there is no mention of any of the post-1977 essays. *Unfinished Synthesis* is more generous.)

7 I have spoken of 'design' on the transcendentalist side. In fact, in the days when natural theology ruled (OK) it was argued that homologies do show design, but everyone agreed that they show a lot less design than functions – and there were those who accepted only functions as evidence of design, who used homologies against design. I speak of the 'crystalline' metaphor after Whewell, who is my window into much of the early nineteenth-century thinking about natural theology (1833b, 1837, 1840).

8 Although I have much admiration for Richard Dawkins' *The Blind Watchmaker* (1986) I think, therefore, that he is quite wrong to dismiss punctuated equilibria theory as a wrinkle on neo-Darwinism. This was true of the first version of the theory, but not of the third version.

9 One final comment on the Gould/Eldredge differences. Given what I have said about Gould's deeper roots in continental philosophy, one might plausibly suggest that punctuated equilibria theory represents more of a paradigm shift for him than it does for Eldredge. This is in no sense intended as a reflection on the relative abilities of the two men, but more on their relative interests. What must also be mentioned is that whereas Eldredge is an ardent cladist, Gould is fundamentally indifferent to problems of systematics. Gould, I see, as a paradigm creator first and a paleontologist second. Eldredge is through and through a working paleontologist – read the love of the subject in *Time Frames* (1985) – and all of his scientific theorizing is directed towards the solution of problems faced by the paleontologist. Punctuated equilibria theory and cladism are first and foremost tools of the trade.

If I am correct in all of this, then the difference between the first presentations of punctuated equilibria theory fall into place. Eldredge (1971) has less panache because it is no more than it seems – proper understanding of Darwinism as applied to the fossil record. Eldredge and Gould (1972), although still within the Darwinian paradigm, is looking for a way out. Gould's D'Arcy Thompson paper (1971), came before the punctuated equilibria papers.

Chapter Six

TELEOLOGY AND THE
BIOLOGICAL SCIENCES

Teleology! Teleology? For two days I've heard nothing but
'teleology'. What is teleology? (Judge William R. Overton,
half-way through my cross-examination at the Arkansas
Creation Trial.)[1]

The most striking thing about animals and plants, separating
them from rocks and lakes and so forth, is that in some sense
they work. Animals and plants – 'organisms' – have an ongoing,
more-or-less independent existence, drawing usable energy from
the surroundings, and in some way or ways reproducing their
kind, so that even if a particular instance of an organism is
destroyed, there are others.

Organisms do not survive and reproduce just by chance. They
do so because their various features contribute to their
possessors' success. Eyes help mammals to see. Fins help fish to
swim. Leaves help plants to attract usable energy. Seeing,
swimming, converting sunlight, these are all things that
organisms do to keep going. Thus you can think of such
processes and the features which make such processes possible,
in terms of the aim of the well-being of organisms. Or, since the
immediate ends of organisms (*qua* organisms) are survival and
reproduction, you can think of processes and features in terms of
their roles in serving the end of survival and reproduction. Thus,
in dealing with the organic world, you can ask what 'function'
or 'purpose' a part or process plays, meaning how does it serve
the end of survival and reproduction. Since rocks and lakes do
not survive and reproduce (as self-subsisting, energy using
entities) you do not ask about the ends served by their parts.

This end-directed thinking and language about organisms is known as 'teleology', although given the conceptual/historical baggage of that term (to be discussed in a moment), some authors today prefer to use a new term like 'teleonomy'. (Mayr (1974) advocates 'teleonomy'; but others, for instance Oster and Wilson (1978), stick with the old terms.) Teleology raises a number of interesting philosophical questions, and I intend to touch briefly on the following in this discussion. First, what exactly does it signify in biology when one uses teleological language/understanding? Or more precisely, what does one think one is talking about when faced with teleology? How is it that teleological language is appropriate? Second, coming out of the first question, is all of biological teleology exactly the same? Third, does teleology mean that in some important respect biological type of thinking is forever different from physico-chemical type of thinking? And fourth, is teleology bad for biology? Should we try to eliminate it, or at least play it down?

WHAT IS TELEOLOGY ALL ABOUT?

Until the middle of the last century, whatever David Hume may have said in his *Dialogues Concerning Natural Religion* (1779), people had little doubt about the significance of teleology. They knew full well what was going on when they were faced with organic features which contribute to their possessor's benefit. As the anatomist Richard Owen (1834) said, when writing of the contrivances by which the baby kangaroo avoids being choked on its mother's milk, there is here 'irrefragable evidence of creative forethought'. In other words, characteristics contributing to their owners' ends – so called 'adaptations' – do so because God has designed them that way. Organisms are, if you like, God's *artefacts*, and just as human artefacts can be thought of teleologically – what purpose does the spout on that receptacle serve? – so also organisms can be thought of teleologically – what purpose does the bump on that lizard serve?

In the case of human artefacts, ultimately they exist for the benefit of humans. In the case of organisms, in the early part of the nineteenth century, there was some debate about whether ultimately non-human organisms served the ends of humans or existed in their own right (Buckland 1836). Organisms which

147

apparently became extinct even before humans first appeared, hardly did much for our well-being. Perhaps organisms' existence and reproduction testified solely to the glory of God. But, whatever the true answer, the basic principle is the same. Teleology is not just the result of blind law. It demands and attests to a thinking creator. (For brevity, I am picking up the story in the years just before Darwin in Britain. I justify this decision by Darwin's great importance to modern biology. But the full story of teleology goes back at least to Aristotle's *On the Soul* and his *Parts of Animals*, and later, on the continent, must refer to Kant (1980). However, these other pre-evolutionary interpretations which did not necessarily involve the Englishman's idea of God, but which had the features I shall identify as characteristic of teleology, support rather than detract from what I shall say.)

Charles Darwin's great evolutionary work, *On the Origin of Species* (1859) changed all of that kind of thinking, at least as far as working biologists are concerned. One can accept his theory of evolution and still be a believer, holding that at some level God was responsible for the design and creation of the organic world. But it was, at best, design at a distance. The immediate cause of organic adaptation for the evolutionist, that which gives rise to the teleological mode of thought, is the mechanism which Darwin made central to his theory: natural selection. More organisms are born than can survive and reproduce. There is a consequent struggle for existence, specifically for mates and for breeding opportunities. Some organisms have, by chance, peculiar characteristics which help them succeed in the struggle. These organisms are thus chosen, or 'selected', as the breeding pool of the next generation. Over time, this ongoing natural form of selection leads to full-blown evolution. And, as a consequence of such evolution, organisms are seen to be adapted, or possessors of adaptations. However, now an adaptation is seen to be a product, not of direct creative intervention, but of straightforward success in the struggle. Those organisms with eyes survived and reproduced. Those without eyes did not. (For a detailed discussion of Darwin and teleology, see Ruse 1979a.)

Darwin's theory is, in modern form and subject to qualifications to be noted later, the dominant view of evolution today

(Ruse 1982c). Natural selection, therefore, seems to be the key which opens the door to the teleological language and understanding in modern biology. When a biologist uses functional language, trying to understand a feature in terms of the ends it serves, although such understanding is *prima facie* the same as that occurring before Darwin, it is now really understanding in terms of natural selection that is at stake (Williams 1966).

IS ALL BIOLOGICAL TELEOLOGY THE SAME?

Historically, we have just seen that the teleology of biology was associated with thought about organisms as God's artefacts. This gives us a clue to answering our second question, namely what about types of teleology. Think for a moment about how a pre-Darwinian would regard the human hand. It is supposed to have been designed and created by God for human benefit. Literally, this work is supposed to have been done by God. But in fact, since we do not have knowledge of such a process, our understanding is a *metaphor* from human experience. We ourselves design and create objects, making the parts to serve the end. Then, we transfer this activity to God, believing Him to have done likewise with organisms. God, like humans, is supposed to have started with an idea or plan, and then to have set about executing that plan, creating objects where the parts would serve some overall purpose.

Is there any other aspect to the human dimension which is teleological, other than that of the artefacts we create? Surely there is, namely the very teleology of our own intentions and purposes and actions. I aim to get a book published. I write a manuscript, and send it off. If it is accepted, all well and good. If it is rejected, I show that I am aiming towards a goal (and my subsequent actions thus become understandable), inasmuch as I revise and resubmit, or try another publisher. I suggest that this is a different teleology from that of artefact teleology, although undoubtedly the two can overlap. (For example, I could decide to make myself into a living saint, and persist in my goal despite all sorts of obstacles like inheritance of a billion dollars.)

The key mark of this second kind of teleology is that it is *goal-directed* (Nagel 1961, 1977). There is a continued aim towards

149

an end, despite disruptions. Do we have a similar kind of teleology (or rather, teleological understanding) in modern biology? I suspect that we do. Organisms frequently exhibit goal-directedness – 'adaptability' – and biologists seize on it, and try to understand what is happening in terms of ends. The most dramatic examples of such teleology occur when there is reason to think (or rather, when biologists think there is reason to think) that consciousness is involved. For instance, recent fascinating reports on chimpanzees suggest that older females act as peacemakers within the groups. Harmony is a desirable state or goal. When males quarrel, harmony is lost. At which point older females spring to life, resolving tensions and quarrels through appropriate grooming and appeasement gestures. Harmony is restored. This sort of behaviour is undoubtedly thought of in a teleological way by primatologists. And there seems to be more at stake than mere artefact teleology. (For details of the chimpanzees, see de Waal 1982.)

There are many clearly non-conscious examples of goal-directedness in nature. For instance, sophisticated mechanisms exist to keep bird-clutch sizes at an optimum (Lack 1954, Ruse 1973c). Such mechanisms are adaptive, and thus teleological in that sense. I suspect that some ornithologists think them teleological in the second sense also. However, this may be a point where biologists disagree (which is just what you expect when you have the application of metaphors). I suspect that some biologists think that goal-directedness is only a necessary condition for extra teleology, and that consciousness is required for a sufficient condition. In which case, chimpanzee behaviour qualifies, but not bird-clutch sizes. Others think goal-directedness alone sufficient, and thus regard bird-clutch sizes additionally teleological in a way that regular adaptations are not. (I would take Waddington (1957) as a member of the first group, Lack (1966) as a member of the second group, and Williams (1966) as ambiguous.)

DOES TELEOLOGY SEPARATE BIOLOGY FROM PHYSICAL SCIENCES?

No one seriously active in science today supposes that the stuff of organisms differs in any significant way from the stuff of inorganic matter (Ayala and Dobzhansky 1974). Proportions are different, of course, but everything quick and dead is made up of molecules, which are in turn made up of the same kinds of atoms, and so on. Moreover, today the physical and biological sciences are increasingly being brought into contact, as concepts and theories from the first domain are applied fruitfully to the second. Arguably the most exciting thing that has happened in biology in the last half-century has been the development of molecular genetics, and its application to virtually every facet of the organic world (Judson 1979).

But what about teleology? Is this something necessarily distinctive of biology, or will it vanish with the successful advance of the physical sciences? My surmise is that there is something distinctive about teleological understanding, and that therefore it will not fade away. Or at least, if it were to fade away, something would be lost. As one who has long championed the influences of physico-chemical models of biology, I do not deny that non-teleological analyses of organic phenomena can be given. I accept, for instance, that much fruitful effort has been devoted to the non-teleological explication of goal-directedness (Sommerhoff 1950, Ruse 1973c). But, I would argue that if it is insisted that, in the future, only non-teleological analyses be given, then something important in our present understanding in biology would be lost.

This seems particularly true of the artefact type of teleology of biology, centring on adaptation. Whenever I read the *Origin of Species*, I am reminded of the wonderful Sherlock Holmes story about Silver Blaze, the stolen racehorse (Conan Doyle 1980). Holmes tells Watson to think about the dog that barked in the night. 'But the dog did not bark in the night!' 'Precisely!' And then Holmes goes on to show that the theft must have been an inside job, because the non-barking dog obviously knew the thief. Analogously, when I read the *Origin*, I am struck by the extent to which the teleology changes from previous non-evolutionary writings. 'But the teleology does not change!' 'Precisely!'

Of course, I exaggerate. As I have argued, in the most important respect, the teleology did change. Adaptation was no longer thought of as God's direct creation, but as the product of the slow process of natural selection. But the teleological cast of mind and language, the way of thinking of features in terms of what they are intended or expected to do – the metaphor of design – changed not one whit. Darwin literally took over the language and imagery of Archdeacon Paley (1802), lock, stock, and barrel. (Darwin had, in fact, studied Paley's works when a student at Cambridge. I suspect he knew of Kant's work, especially through his close relationship with Whewell (Ruse 1975b).)

This, it seems to me, points to something very important. There is a flavour to the phenomenology of the biological experience which transcends the non-evolutionary/evolutionary divide. Whatever may really be the case, organisms seem *as if* they were designed. Why this condition had to be so is a hard question; although, I suspect Hume (1779) was on the right track, when he argued that a thing could not actually get to work as a living organism unless it were complexly integrated in a quasi-design fashion. Be this as it may, the point about the organic world is that, unlike the rest of the universe, it has this artefact look about it. This is what the teleological approach in biology seizes upon and exploits, and why the teleological approach is appropriate. However, the other side of the coin is that if you were to drop the teleology, then you would do so only by ignoring the design-like character of living matter. And this would be to ignore and lose one of the most important things about organisms (see also Beckner (1969) on this point).

I doubt, however, that there is much danger of this loss happening. As the physical sciences have moved into biology, it is they who have had to do the accommodating! There has been no question of eliminating teleology. Rather, the new molecular biology has taken it up with vigour. The notion of a genetic code is as artefact-like as anything in traditional biological science. Molecular genetics emphasizes the distinctive, end-fixated nature of the organic world. It does not deny it.

TOO MUCH OF A GOOD THING?

I trust now that none of my readers harbour secret thoughts of eliminating the teleology of biology. The design-like nature of organisms is as much a part of them as is their molecular composition. But is this 'design' really as far-reaching as some have supposed? I refer now not only to old-fashioned natural theologians, who see God's hand everywhere, but also to many modern Darwinian evolutionists, who see all-pervasive the effects of natural selection. Some critics argue that, as Darwin himself sometimes admitted, in biology teleology tends to get over-emphasized (Gould and Lewontin 1979). These critics feel we should not expect to see every aspect of organisms functioning towards the overall benefit of the organism. It is more reasonable to suppose that many features serve no end at all. (Today, other than in rather special circumstances, few think benefit is to the group of unrelated individuals – it is always to the individual organisms, or to close relatives.)

In reply, I would say three things. First, no one claims that every aspect of the organic world serves an identifiable purpose (Dobzhansky *et al.* 1977). I doubt that the human species would come to an abrupt end if males lost their nipples. Second, one should be wary of categorical claims that features could not possibly have a purpose. Philosophers have long delighted in taking heart sounds as having no value, but in fact they stimulate infant suckling and bonding (Forbes and King 1982). The four-limbedness of vertebrates, to take another example of an oft-cited supposedly purposeless phenomenon, likewise is readily connected to plausible adaptive functions (Maynard Smith 1981). Having four limbs, two fore and two aft, helped the first vertebrates, fish, go up and down in the water. (There is no claim by today's theologists that that which had an immediate purpose always helps that very purpose.) Moreover, recent supposed evolutionary rivals to Darwinism, like punctuated equilibria, although purportedly playing down adaptation and function, prove nevertheless, on close examination, to have a large role for end-directed understanding. No one has yet denied that the hand and the eye and feathers have purposes. (Gould, the most vocal advocate of punctuated equilibria theory, is much more open about adaptation in his more recent writings. Compare him now to Gould 1980b and 1982b, even. Other proponents of non-Darwinian mechanisms, like Kimura (1983), give adaptation a full place at the non-molecular level.)

Third, we must not forget that pushing the teleological mode of understanding pays huge heuristic dividends. Biologists devise 'optimality models' (pictures of how, in theory, organisms might best be expected to use their resources) and then can check these against nature. Time and again such models, based on the notion that organisms are the most efficient possible artefacts, have proved their worth (Maynard Smith 1978c). Recently, for instance, such models have thrown bright light on the use of resources in social insects – why, for instance, the ratios of certain castes of workers and soldiers are what they are. Without such models we should have remained ignorant (Oster and Wilson 1978).

In short, the teleology of biology seems here to stay, and judging from the use biologists make of it, that is no bad thing.

NOTES

1 Note added in 1988: see the introduction to Section IV for the background to this trial.

Part III

HUMAN PERSPECTIVES

'Light will be thrown on the origin of man and his history.' We have seen that right from the moment he became an evolutionist, Darwin never for one moment doubted that we humans are part of the evolutionary picture. And the very first record we have in his private notebooks that he had unambiguously grasped selection focused on the mechanism's importance for the development of human reasoning powers.

Yet despite the fact that in the *Descent of Man* of 1871 Darwin explored in detail the implications of his thinking for human-kind, for a century this part of the paradigm languished, not thrived. There were various reasons for this. One was the innate difficulty of understanding so complex an organism as *Homo sapiens*, another the rise of alternative (secular) world-pictures like Marxism and Freudianism. However, in recent years Darwinism has started to move forward, thanks particularly to the ideas and discoveries of the students of animal social behaviour: the 'sociobiologists'. (For a defence against contrary claims that sociobiology is indeed part of Darwinism, see Ruse 1987a.)

Human sociobiology has been, and still is, dreadfully contro-versial. As I explain in the first essay of this section, the subject can only be judged on its real achievements. So what I have tried to offer here is a flavour of precisely such achievements. To me the most important marker is the work of the young anthropologist Monique Borgerhoff Mulder, on the Kipsigis, in which she tries to show the importance of biology for an understanding of marriage patterns. As always, a good way to see if a science is forward-moving is to find out if bright students are being attracted to the field in question. It is their careers which are at stake.

Yet can one cavalierly dismiss the objections to human sociobiology? It has been accused of racism, sexism, capitalism, and more. Simply ignoring such charges hardly makes them go away. My own feeling is that, as much as possible, one should respond to such charges positively, admitting faults where they exist, trying to correct errors, and all of the time digging for deeper understanding. Such at least is my aim in the second essay, on the supposedly sexist nature of Darwinian thought, past and present. That such has been there can hardly be denied. Nor is it an excuse to point out that everyone was a sexist until about 1975. Nevertheless, as we try to mend our ways, we can learn more of science itself and of Darwinism in particular. Responding to a stimulating thesis about the status of values in science, by the Catholic historian and philosopher of science Ernan McMullin, I argue that one can preserve the objectivity and progressive nature of Darwinism while allowing that values (of all kinds) will be with us always. The trick is to make sure that the values in our science are the right values.

And this brings me to the essay on extraterrestrials, one of my own favourites, even though it is not really on extraterrestrials at all! This is not to say that I think life on other worlds a silly subject or not worthy of philosophical attention. Indeed, as you will see, I much regret that today's philosophers (unlike those of the past, such as Darwin's very serious mentor William Whewell) avoid the subject, for it is a terrific topic from which to launch into a myriad of different issues. In my own case, I speculate about hypothetical Andromedans to see just what implications my commitment to Darwinism has for non-hypothetical humans.

Of special concern to me here are the traditional questions of philosophy: What can I know? What should I do? Most people in my field would react rather negatively at this point, thinking that a scientific theory like Darwinism can tell us little or nothing about these crucial matters, and that to suppose otherwise is to commit all sorts of errors, like the so-called 'naturalistic fallacy'. Until recently, I myself would have agreed with such sceptics, but now I have swung around nigh 180 degrees. I think that the fact that we are modified monkeys rather than the special creation of a good god on the Sixth Day must speak both to epistemology (theory of knowledge) and to ethics (theory of morality). My only regret is that in this essay I did not push my new-found liberation farther than I did. In the final essay of this collection, I will do just that.

Chapter Seven

HUMAN SOCIOBIOLOGY: AN INTERIM REPORT

Our point of reference has to be Edward O. Wilson's *Sociobiology: The New Synthesis*, published in 1975. In that work he offered a comprehensive survey, theoretical and empirical, of a burgeoning subject he labelled, 'sociobiology': the study of animal social behaviour from an evolutionary (more precisely, in Wilson's case, from a *Darwinian* evolutionary) perspective. Although much praised, the efforts of Wilson and many of his fellow workers were also much criticized, primarily because he (and others) insisted that an evolutionary perspective throws light, not only on the dumb animals, but also on the thought and action of us humans (Ruse 1985, Montagu 1980).

And so, history repeated itself. In the last century, there was the Bishop of Oxford and his friends berating Darwin and his companions, in public debate and through such influential organs as the *Quarterly Review* (Ruse 1979a). In this century, there was fellow-Harvard-biology-department-member Richard Lewontin and his friends berating Wilson and his companions, in public debate and through such influential organs as the *New York Review of Books* (Segerstrale 1986).

It is too soon to write a history, but an interim report is surely in place. So, I intend to look at the state of play in human sociobiology, as it stands some ten or so years after Wilson's *magnum opus*. My emphasis will be on positive achievements and prospects, partly because I have elsewhere looked at criticisms, partly because there are others far more eager than I to look negatively at human sociobiology, and partly (mainly) because of my long-held view that much of the criticism is irrelevant, anyway (Ruse 1979b).

159

No doubt Popperians – not to mention those who have labour-ed long in the cause of truth, however stern and uncomfortable – will deplore my last point. But, as a good evolutionist, I am strongly influenced here by the past. The theory of Darwin's *Origin* was nigh overwhelmed with problems, internal and external, and there was no shortage of voices happy to proclaim this fact loud and clear. One thinks, for instance, of Darwin's lack of an adequate theory of heredity, and of his troubles with the age-of-the-earth question (Vorzimmer 1970, Burchfield 1974). But Darwin succeeded because (in Thomas Kuhn's language) he gave biologists a good, working paradigm, within which they could study life's problems. Without the *Origin*, H.W. Bates and A.R. Wallace could not have worked on butterflies, nor could J.D. Hooker have made his brilliant findings about plant distributions (Ruse 1979a). Likewise, I believe that human sociobiology will succeed or fail on its successes, or lack thereof. Does the study of human thought and social behaviour, from a Darwinian perspective, lead to new insights and findings, or not? If it does, then criticism (however well-taken) will be secondary. If it does not, then criticism is not needed.

Enough of apology for what I am not going to do. To the task at hand. My discussion begins with background remarks about biology generally and human biology in particular, trying to set human sociobiology into proper context. Then I turn to claims by people active, at various levels, within the field. Preferring depth to breadth (although, alas, not great depth), I shall foreswear an overall survey and shall concentrate on two specific examples, one rather broad and one rather narrow.

DARWINIAN THEORY

Looked at from afar, a good date to start our story is 1859, rather than 1975. It was then that Charles Darwin published his *Origin of Species*, wherein he argued that all organisms (including ourselves) are the products of a long slow process of evolution through 'natural selection'. More organisms are born than can possibly survive and reproduce. There is a consequent 'struggle' and the victors (the 'fitter'), who tend on average to have heritable features not possessed by the losers, are thus 'picked' or (in a process analogous to that done artificially by the breeder) 'selected'. Given enough time, this leads to full-blown evolution. But (and Darwin, who had been much influenced by natural theology, was most insistent about this), the process does not lead merely to change, with organisms coming out in any old way. The end result of evolution is functioning entities, with organisms showing 'adaptations', like hands and teeth and ears and eyes – features which aid their possessors in the struggle. It must be added that it was Darwin's life-long conviction that such adaptations essentially serve only their possessors. Something could never be selected purely for the benefit of some other organism – in today's terminology, Darwin was ever an 'individual selectionist' rather than a 'group selectionist' (Ruse 1980a).

Darwin's ideas, both in general and in particular, have persisted. For instance, after years of neglect and denial, his views on the significance of individual selection have been re-recognized. However, naturally enough, his theory has been modified and augmented, most particularly by the incorporation of ideas about the nature of heredity. As is well known, in recent years these ideas ('genetics') have drawn heavily on the theories and findings of the molecular biologists. I shall not stop to debate Darwinism (or neo-Darwinism), stating simply and categorically that I believe it to be a good tough theory, which properly can and does serve as the basis of our thought about organic origins. This does not mean that it is complete or perfect – it certainly does not mean that every last item is produced and tightly controlled by selection – but it does mean that, in outline, it is true (Ruse 1982c, Ayala 1985).

Moreover, it is still growing. Most exciting for evolutionists in the past two decades has been the way in which Darwinism has been extended to the study of animal social behaviour – to the subject matter of 'sociobiology'. Right through the animal kingdom, biologists have been showing that what animals do, particularly in relation to each other, is as crucial to their biological success as is any feature of their physical nature. Behaviour is a function of the genes as sorted by natural selection, and it succeeds and is preserved because it confers adaptive advantage on its possessors (Trivers 1985).

What of our own species, *Homo sapiens*? As even Darwin's own critics soon recognized, if you are in any way faithful to the spirit of science, you must agree that we humans are part of the natural world, and that we share ancestors with the other denizens of this world. The extraordinary similarities (homologies) between us and other mammals speak more than could any number of words. It is true that, sociobiology aside, the working out of the implications of human evolution have proven long and arduous, and are hardly complete even today. Owing to hurdles ranging from lost evidence to outright fraud, only very gradually have the mists started to rise from that path which leads into knowledge of our past. But that this path lies through the valley of evolution can be denied by no reasonable person (Isaac 1983).

Most excitingly, in recent years, the mists have started to disperse at an increasing pace. Thanks to a much-improved fossil record and to powerful new molecular techniques, we can at last say many significant things about our ancestry. Quite incredibly – certainly, incredibly judged by what was believed a few years ago – it now appears that the proto-human line broke from that of our closest relatives (the chimpanzees) only six million years ago, with the proto-gorillas hiving off some three million years earlier. To put matters in perspective, the age of mammals began 60 million years ago, the first mammals appeared 200 million years ago, and life started over three and a half billion years before the present (Pilbeam 1984).

Hardly less exciting is our realization that of the two most distinctive human features – bipedalism and a big brain – the one (bipedalism) was virtually achieved (by four million years before present) before the other began its growth (from 400 cc) to its present state (1,400 cc). In itself, this is not directly a causal fact, but it has immense implications for our thinking about the forces behind human evolution (Johanson and Edey 1981). Cutting short a long and still-developing story, I shall say simply that although, obviously, no one was around to watch causal forces in action, all that we know suggests and confirms that the major factor in recent human evolution was natural selection. For instance, the capacity for tool use has obvious adaptive virtues, and there is a close correlation between the sophistication of such tool use and the growth of brain size.

Where stands human sociobiology in all of this? The claim of its enthusiasts is analogous to that of the animal sociobiologists: evolution is still with us, in spirit and in fact. The only way to understand humans is as products of our evolutionary past, and as processes of our evolutionary present. We are objects produced by the units of inheritance (the genes) as sorted and selected in the struggles for survival and reproduction. Human thought and behaviour – particularly, human *social* thought and behaviour – results from and must be related to biologically rooted adaptive advantage (Wilson 1978, Alexander 1979).

Thus, without going into details (for some specific instances will be taken up shortly), we find that the sociobiologists are much interested in family relationships, and how various moves and counter-moves might be adaptive, that is directed towards reproductive success. Again, sociobiologists care about the ways in which humans deal with their non-related fellows. Why are we so friendly at times, and why does such friendship occasionally break down, even exploding into all-out conflict? Can the madness of this century really be a product of our animal past? Did evolution lead straight to Ypres and the Somme? Did it lead to Auschwitz and to Hiroshima?

The light which guides the human sociobiologist is natural selection, culminating in genetically based adaptive advantage. In line with Darwin and his other followers today, the emphasis is much on individual selection. There may be – there surely is – much co-operation in the human world, but ultimately the benefits have to return to the individual. (Curiously, the one human sociobiologist who is prepared to take seriously non-individualistic mechanisms of change is Wilson. This stems from his holistic philosophy, symbiotic with his work as the world's leading expert on the social insects. The complexity and integration of life in that social realm inclines him away from totally orthodox Darwinism.)

There is much that could be said about human sociobiology. With an eye to future discussion, to round out my background remarks I shall pick up and elaborate on two crucial points.

OF BIOLOGY AND COOKERY

The first point focuses right in on the intentions of the would-be human sociobiologist. In principle, and indeed in practice, there seems to be a range of possibilities. (My thinking on these questions has been clarified by Borgerhoff Mulder 1987b, and

Caro and Borgerhoff Mulder 1986.) At one extreme, one might take a hardline, so-called 'biological determinist' stance, believing that the genes tightly control behaviour, no matter what, why, or when. My eyes are brown, because of my genes. My profession is philosophy, because of my genes. I suppose that such a stance might be taken for metaphysical reasons (*i.e.* you believe that it really is true), or methodological reasons (*i.e.* you believe that, at this early stage of sociobiology's development, you simply must and should concentrate on the easiest problems, and these are yielded if you assume a tight connection between genes and behaviour).[1]

I know of no one who is consistently a hardline determinist, metaphysical or methodological, despite the fact that popular present accounts of human sociobiology are frequently adorned with cartoons of clockwork human figures (with keys protruding from their backs) or of marionette men, dangling on strings controlled by the double helix. Metaphysically, no biologist worth his or her salt would deny that all organic features, most especially behavioural features, are a function of the genes interacting with the environment, which latter might change drastically, with no less drastic effects. Methodologically, no one would think that ease of modelling was worth quite so high a price, as never to admit that the genes are but one of a cluster of causal factors.

However, it is certainly true that some human sociobiologists would incline more to the deterministic end of the scale than others. I think this is true of Wilson, who suggests that culture can have a feedback effect on the genes in a very short time, (especially in the overall evolutionary perspective, Lumsden and Wilson 1981, 1983). Yet I do know that much of Wilson's motivation *qua* scientist is more methodological than metaphysical. He believes the only way to make problems 'tractable' (a favourite word) at this stage is by devising much simplified models, ignoring complexities of gene expression. He points to the history of genetics in particular and of science generally, as his guide and support. T.H. Morgan and his associates in the Columbia fruit-fly room did not begin with the more complex aspects of Drosophila development. No more should the human sociobiologist begin with the more complex aspects of human behaviour.

Right at the other end of the scale from the determinist there may be someone who would decouple biological nature and change, and human cultural nature and change (that is, the realm of human thought and behaviour), virtually entirely. Such a person will grant biology a place in our basic animal needs – eating, drinking, copulating, defecating, and the like – but would argue that the way we do these things, and all else, lies outside biology. There may be similarities between biological and cultural change – there are undoubtedly differences – but they are separate processes. (This is close to the position of Durham (1976).)

I am not sure that such a person as this should really be called a 'sociobiologist', although I (for one) would certainly not demand the doctrinal uniformity required by most Christian sects. (In this respect my ideal is the Church of England!) But I shall have little to say about such a position, in theory or in fact. What is of real interest to me at this point is the large number (surely the majority) of human sociobiologists who fall in the centre, between extremes. They believe that the genes have an important input into human thought and behaviour. They believe that such thought and behaviour reflects Darwinian adaptive strategies (that is, *biological* advantage), no less than hands and eyes. They also believe that the connection between genes and behaviour is complex and significant, and must be acknowledged right from the beginning. This, I would add, is the group within which I would wish to be counted.

Developing this centre-line sociobiological position, we might avail ourselves of a happy metaphor which has been proposed by the biologists Patrick Bateson and Richard Dawkins. They liken the connection between an organism (including its thought and behaviour) and its causes to that between a cake and its recipe. Bateson writes:

Is it really possible to break up the fully developed song of an experienced male chaffinch into components, some of which are specifically affected by experience and some of which are not? Even though we know that many factors have been responsible for the detailed specification of the song (Thorpe 1961), it does not follow that somehow these factors will correspond to the constituents of the final behavioural product. Rather than liken the development of such behavior to the insertion of days into an existing calendar (*intercalare*), I suggest a more appropriate analogy would be the baking of cake. The flour, the eggs, the butter, and all the rest react together to form a product that is different from the sum of the parts. The actions of adding ingredients, preparing the mixture, and baking all contribute to the final effect. The point is that it would be nonsensical to expect anyone to recognise each of the ingredients and each of the actions involved in cooking as separate components in the finished cake. For similar reasons, I think those cases in which a simple relationship can be found between the determinants of behavior and the behavior itself will be exceptional. (Bateson 1976)

Likewise Dawkins:

The genetic code is not a blueprint for assembling a body from a set of bits; it is more like a recipe for baking one from a set of ingredients. If we follow a particular recipe, word for word, in a cookery book, what finally emerges from the oven is a cake. We cannot break the cake into its component crumbs and say: this crumb corresponds to the first word of the recipe; this crumb corresponds to the second word in the recipe, etc. With minor exceptions such as the cherry on top, there is no one-to-one mapping from words of recipe to 'bits' of cake. (Dawkins 1981)

Of course, this still leaves you with precise questions about development, and how particular aspects of thought and behaviour might be produced, especially since there is more to life than cherries. Unpacking the cake metaphor, I would remind you that although the eggs in a cake do not (should not) make the cake come out tasting 'eggy', they do have a definite effect – what is surely an 'adaptive' effect – namely that of making the cake rise. No eggs – no fluffy creation that melts in your mouth. However, within these bounds, there are all sorts of cakes that might be produced.

Perhaps the genes affecting human behaviour have a like effect: they set certain constraints, rooted in adaptive advantage, but the actual expression of thought and behaviour is not laid down precisely. I will say no more here about the possibility and nature of such innate dispositions (as much dispositional as innate), but shall return to the idea later. (I will also ignore obvious side issues, such as the existence of eggless cakes and the crucial importance of correct oven temperature in producing risen cakes. My suspicion is that these work for the metaphor, rather than against it.)

One final matter and the arguments of this section are complete. The human sociobiologist is certainly committed to the view that human thought and behaviour were shaped by natural selection. Is the presumption that selection is still operating and (more crucially) that humans, particularly those living in industrialized societies, always think and do that which is of immediate adaptive advantage?

In an age of AIDS and Chernobyl, I doubt that anyone would argue for so tight a link between behaviour and adaptive advantage. But, certainly some (Wilson?) would see adaptive advantage tracking environmental change fairly closely. Conversely, none, I suspect, would want to say that adaptive advantage is totally irrelevant, for even the most modern of people. It should be noted, however, that one does not have to be an extreme determinist to think adaptation important. One might, for instance, think that the genes promote a kind of flexibility, so that one can maximize adaptive advantage, even as the environment changes. (To illustrate this point by analogy, consider two machines, both designed with the aim of going from a point A to the nearest of a set of points B. If B is the closest B, one can imagine one (determinist) machine programmed to go to precisely B whereas the second (non-determinist) machine has the flexibility to focus on any other B, should it move closer to A than B. I suspect most sociobiologists think we are more like the second machine, although few would think we are quite as exactly controlled by selection.)

We shall return again and again to the matters covered in this section. But for now, they should be clear enough to stand on their own. Let us therefore go on to the second promised point of clarification and elaboration.

OF PARADIGMS OLD AND NEW

The philosopher David Hull (1978a) has remarked sensibly that names are powerful but dangerous things with lives and forces quite of their own. The label 'sociobiology' (more specifically, 'human sociobiology') served to identify and consolidate a field, making something which (for instance) could have its own specialized journals and organizations. It also gave the critics an easily demarcatable target. The overall result is that the subject is too-readily seen (by both supporters and critics) as an isolated, entirely new area of science. The wagon train has been drawn in a tight circle, and too many think that what lies within is a world entirely unto itself.

This is a gravely distorting picture of the way things really are. Human sociobiology does not exist in splendid isolation, breaking entirely new ground, and earning all the credit and blame for itself. Sociologically, it is probably true that human sociobiology does rather function as a Kuhnian paradigm, with supporters rallying round the flag and detractors sniping (sometimes letting off heavy cannon) from without. (I will pick up on this point, later.) But, epistemologically, human sociobiology must be seen as part of the overall Darwinian synthesis (Ruse 1987a). It is a branch of basic evolutionary theory, no less than is paleontology, or biogeography, or systematics. Furthermore, our understanding of the particular contingencies of our own species cannot and should not be divorced from what other biologists tell us about ourselves, for instance from what those who study our recent past (the paleoanthropologists) have to say.

What does this mean, in effect? It means that human sociobiologists have an obligation to speculate and a certain credit to draw on. We know that natural selection was crucially important in the development of life. We know (at least, we have strong evidence that) natural selection was important in the development of human life. We know that evolution through selection is (what I like to describe as) a 'string and sealing wax' process. You do not start anew in each generation or with each species. Rather, you pick up and mould and twist and improvise, with the successes of the past being used again and again, either as they were first intended or for some new purpose. We know (and this is a point that I merely mention without discussing in this paper), that animal sociobiology strikes new successes, almost daily. Moreover, some of the greatest successes have been with primate species, close to ourselves. Overall, therefore, the presumption is that sociobiology might throw some light on us humans. To do this, they must (and may) speculate and expect to be taken seriously (at least, genuinely tolerated).

I am not making a plea for hypothesis without evidence, nor do I want to be trapped in a circle, where speculation is taken as fact, which is then taken as support for further speculation. Human sociobiologists had better come up with some findings and some predictions and some connections, or we shall start relegating them to the status of parapsychologists – dreadfully sincere but dreadfully implausible. But at present human sociobiologists do have some leeway – at least, this is what they should be offered. And, this is especially true in the light of what one might euphemistically describe as the lack of success of much conventional social science theorizing in this century. If you are really sincere about a scientific understanding of humankind, you had better not strangle a newcomer at birth, especially when that newcomer has influential parents.

Perhaps I might illustrate the point I am making with an example, namely that of homosexuality (especially male homosexuality) (Ruse 1988a). Human sociobiologists have had a number of things to say about this, and have come up with a variety of suggestions as to its evolutionary basis. Best known (perhaps, most notorious) is the hypothesis that male homosexuals play a role akin to worker ants, where they selflessly forego reproduction on their own account, in order to help out their families. They are, to put the matter crudely, overgrown hymenoptera (Weinrich 1976).

Now taken on its own, this hypothesis strains credulity; more bluntly, it is ludicrous. So much so, in fact, one starts to wonder if the sociobiologists are out to denigrate a vulnerable minority, as if today's homosexuals do not have enough trouble with AIDS and recent US Supreme Court interpretations of the Constitution, they have also the suggestion that they are genetically manipulated insects, serving the ends of others.

However, when you look at the full context, the sociobiological speculations (and I will call them no more than that) are nowhere like as grim or absurd. On the one hand, human sociobiologists are Darwinian evolutionists, so they expect reproduction to be a (to be *the*) crucial end for any organism. (This is not to say that it is an end always consciously sought.) On the other hand, doing what homosexual inclinations lead you to do, seems not to lead to babies. Hence, there is a paradox which it is incumbent upon human sociobiologists to explain. They would be lax were they not to do so.

But, why the worker ant hypothesis? Again, the overall context makes things clear. The greatest triumph of animal sociobiology is W.D. Hamilton's (1964a, 1964b) explanation of a phenomenon which had troubled Darwinians since the *Origin*, namely the sterility of the worker castes in the ants, the bees, and the wasps. He was able to show that such sterility follows from natural selection (in the form known now as 'kin selection'), because of the rather peculiar breeding methods of the hymenoptera. Briefly, while females have two parents, males have only mothers. This means that females are more closely related to sisters than to daughters, and hence selection favours the raising of the former over the latter. There is reproduction by proxy, as it were! (see Figure 7.1).

Humans are not hymenoptera. Both sexes have two parents. Moreover, human homosexuals are in no way physiologically infertile. Yet, given the importance of kin selection with the hymenoptera, and its result in non-breeding, it is at least worth an analogous hypothesis when we come to human homosexuals. It might not work out, it probably will not work out, but the scientist who never takes a risk – never makes 'bold conjectures', to use a familiar phrase – is the scientist who has quit doing science.

Of course, you must go on from here. Do homosexuals really have reduced fertility? Perhaps they are supersexed, and can service both males and females. Do homosexuals aid the reproduction of close relatives? In our society, they are frequently pariahs, the last thing a self-respecting family wants in its circle. And, what about proximate mechanisms? What makes someone a homosexual, whatever the evolutionary payoffs?

As it happens, sociobiologists and co-workers have started to answer some of these questions. There is some evidence of reduced reproductive accomplishments (Ruse 1988a). There is

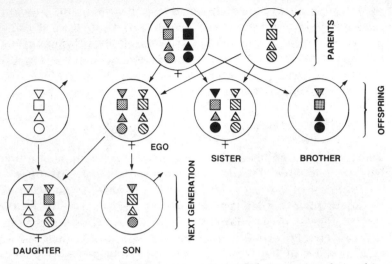

Figure 7.1 A diagrammatic representation of the genetic relationships in the Hymenoptera.

Females are diploid; males are haploid. Only females have fathers. It can be seen that sisters have a 75% shared genetic relationship, whereas mothers and daughters have only a 50% shared genetic relationship. Kin selection therefore favours the raising of fertile sisters rather than fertile daughters. Males have no such special relations, and therefore do not form sterile worker castes.

some evidence, in pre-literate societies if not in our own, that homosexuals do (at least indirectly) lift up the overall family status (Weinrich 1976). For instance, shamans are often exclusively homosexual, and yet they can hold great power and accumulate high benefits for family members. And there is some evidence of links between underlying biology and overt behaviour. Hormonal and other studies are pinning down proximate causes in ways that (at a minimum) synchronise with sociobiological work.

The case is hardly complete, but it is far from empty or stupid, so long as you take the overall perspective. At the very least, given the appalling quality of much of the research into homosexual orientation and behaviour, it merits some opportunity to make its point. Although mention of already-existing research brings out a point which, I suspect is often true – or, at least, should be true. The sociobiological explanation need not conflict with

171

explanations in other fields, like psychology or (as in this instance) endocrinology. All can be and should be working together, some at the proximate causal level, some at the ultimate causal level, and some at the interface. Human sociobiology need not present the conventional researcher with an 'either/or but not both' position.

SIBLING INCEST BARRIERS

There has been enough said by way of prolegomenon. The time has come to look directly at human sociobiology in action. I begin with a broad example; one which appears to be accepted (in some form) by all human sociobiologists. I refer to the sociobiological explanation of the incest barriers which exist between closely related human beings, specifically those which exist between siblings (van den Berghe 1983).

Human beings like to reproduce – they certainly like to go through the motions that lead to reproduction – and, when they are thrown into close and prolonged contact with members of the opposite sex, they are wont to do so. Yet, at the most readily available level, namely that between siblings, sexual intercourse tends not to occur. Affection can and (after a fashion) usually does exist, but it is not erotic. Of course, sex does sometimes occur, and there are even cases where offspring are produced. But sex and reproduction is not common, in fact is miniscule compared to what could have been expected.

Why is there this disinclination towards sibling incest? Anthropologists, like Claude Levi-Strauss (1969), suggest that it is rooted in our culture, and serves the end of forging alliances between groups: sisters can be married off to rival tribes. Freudians give a Lamarckian explanation, finding its beginnings in some patricidal blood act against the father and the resolve of the brothers to avoid competition between themselves (Freud 1913). Sociobiologists take up an explanation first mooted by the nineteenth-century anthropologist, Edward Westermarck (1891). They argue that our biology turns us off: thanks to our genes, as selected in the struggle, sisters or brothers make up the one group from which we do not choose our sexual partners.

Why do sociobiologists argue as they do? There are a number of reasons, the strongest of which undoubtedly is that close inbreeding has appallingly deleterious effects. Offspring of siblings are all-too-frequently horrendously genetically handicapped (Seemanová 1971). This means, therefore, that down through the ages there would have been strong selective pressures against sibling incest. Hence, what better biological adaptation to prevent such incest than a natural disinclination towards this kind of sex?

Backing this central argument is a range of empirical findings. Highly significant is the fact that, however varied cultures may be, with some few exceptions (to be mentioned below), incest barriers are found in place. The Greeks did not reproduce with their siblings. No more do we, or the Yanomamo (pre-literate people from the Amazon region). Such lack of variability is not absolute proof that the genes play a significant causal role; but it is a strong point. (It is worth noting that when critics attack human sociobiology, cultural variability is a favourite weapon. The human sociobiologists turn this argument on its head, claiming that cultural stability is a confirming phenomenon.)

Incidentally, given that there is this lack of variability across time, as well as across contemporaneous cultures, we have here a point at which human sociobiologists feel emboldened to argue that adaptive advantage persists, even into modern society. We have seen that there can be no guarantee that what was advantageous in our past is adaptive today in modern society. However, since genetically handicapped children are usually at a disadvantage even in modern society, here the sociobiologists see continued adaptive advantage to the natural disinclinations towards sibling sex. This is not to say that the peculiar pressures of modern life might not distort these disinclinations. Rather, that inasmuch as they persist, they have much the same biological value.

An equally significant line of support comes from the animal world. Humans are not chimpanzees, neither are they codfish. Nevertheless, one would expect to find incest barriers in the animal world. Extreme inbreeding should affect other species members no less than humans. Hence, there should be obstacles to the success of sex between siblings.[2] Of course, in cases where there are natural (almost contingent) reasons for dispersal, there is little need of avoidance mechanisms by individual organisms. But where there is need, one expects such mechanisms.

And one finds them. In a survey spanning the mammalian and the avian worlds, Paul Harvey and Katherine Ralls (1986) found barriers in species after species. Even studies sometimes cited as demonstrating high-incest levels are found, on re-examination, to point to this conclusion. Acorn woodpeckers (*Melanerpes formicivorous*), prairie dogs (*Cynomys indovicianus*), and (surely significant for us) chimpanzees (*Pan troglodytes*), all choose their mates from outside the family. Most impressive is the very peculiar mouse-like marsupial genus *Antechinus*. The males all die, at the age of one year, a few days after the mating season. There are, therefore, no adult males when the females give birth, so that there is nothing external to stop male siblings staying home and copulating when mature. But they do not. All of the males go off, in search of mates elsewhere. 'The only adaptationist interpretation left seems to be inbreeding avoidance' (Harvey and Ralls 1986, p. 576).

Of course one might accept empirical findings like these, and yet argue that human incest avoidance has reasons other than those which lead to like barriers in the non-human world. It would not be enough simply to say that humans have taboos (which are cultural) whereas animals have barriers (which are biological), for not all humans do have taboos; unless one can give plausible reasons, one's negative stand starts to look increasingly *ad hoc*. And one's case is made even weaker if one ignores or plays down yet another piece of evidence, namely that there are now some clues towards the proximate mechanisms of human (sibling) incest avoidance (Wilson 1978, van den Berghe 1979, Ruse and Wilson 1986).

Perhaps, expectedly, it is not that we are simply sexually 'turned off' biological siblings. Presumably, such a mechanism as this would require some sort of sensing, through smell and/or pheromones or some such thing. However, humans are generally rather insensitive to channels like these. The true story seems to be that we go though a kind of negative imprinting, being sexually indifferent to those with whom we spend our childhood, or those with whom we 'share the same potty', in one anthropologist's phrase. If you grow up with people through the crucial years between two and six, you do not desire them as mates in adult life. This fact is confirmed by study of a number of natural occurrences where biological siblings have been separated at birth or where non-biologically related children have been brought up as if siblings. The best-known instance of the latter kind of situation occurs in the Israeli kibbutzim. Non-related children reared as social siblings simply do not want to mate. This is despite opportunities to do so (and the lack of any legal sanctions), and despite the fact that unusually close bonds tend to exist between the children of the kibbutz (Shepher 1979).

PROBLEMS

These are some of the positive pointers towards the sociobiological case. What about the other side? What can be said against the claim that sibling incest barriers are rooted in biological nature, adaptively preventing close inbreeding? For a start, there is the fact that we are dealing only with *sibling* incest. What about parent-child couplings, which ought biologically to be just as disadvantageous? I take it that someone who raised this point would be doing so less as an outright criticism and more as a complaint that the sociobiological attack on incest is incomplete. And I would agree that sociobiology would be seriously incomplete were there no evidence of biological barriers to parent-child incest, or if (far worse) parent-child incest were common.

But of course there is some evidence and it is not common. This is not to say that parent-child incest never occurs, or that modern industrial life may not set up special pressures or opportunities that make it more common than before. Nevertheless, intra-societal studies confirm that parent-child incest is not a regular human practice (van den Berghe 1979). At least, it is much less common than what one would expect among unrelated adults and children. (Hence in fairness to sociobiology, I think one should exclude step-parent and child relationships.) It is also of note that the Harvey/Ralls survey of animal barriers applies as much to parent-child relationships as it does to sibling relationships.

The big query (with respect to parent/child barriers) hovers over proximate mechanisms. Presumably, children are open to negative imprinting against regarding their parents sexually. If there is any truth in what Freud writes, such imprinting is fraught with ambiguity and tension – although, supposedly, problems

175

stem less from failure of Oedipal barriers than from guilt at wishing the barriers would fail. Something else other than childhood negative imprinting must be found for parents, expecially fathers. (Some fairly obvious reasons spring to mind for the lack of mother-son relationships, beginning with those rooted in physiology.) However, what this all points to is surely only a gap – albeit a significant gap – in sociobiological thought. The need to explain grown men's general disinclination towards incest with their daughters hardly undermines the whole sociobiological approach towards incest barriers.

Another problem with (or another gap in) the sociobiological account of incest obviously lies in the exceptions. What is one to make of exceptions such as brother-sister matings among ancient Egyptian royalty and nobility, or, of similar cases in Hawaii and elsewhere? Special explanations have been offered for these exceptions, usually based on the virtues of keeping power and property within the family (van den Berghe 1983); and vigorous criticisms (from within and without sociobiology) have been counter-offered (Kitcher 1985). Without wanting to appear cowardly, I shall here side-step debate. Explanations must be sought for the exceptions, that is true, and sociobiologists must not stop until the explanations are adequate. Nevertheless, the exceptions hardly destroy the sociobiological case.

Virtually no claim in physics and chemistry, and absolutely no claim in biology, is without exceptions (Ruse 1973c). In some cases, you could drive a coach and four right through. For example, Boyle's law holds for no gas exactly, and at most temperatures and pressures not at all. Yet the exceptions are no reason for giving up science. Boyle's law is a fundamental part of basic physics. In science, one works step by step, first mapping out basic points, then going on to tackle exceptions and to refine the original ideas. From Boyle's law to van der Waal's equations to ... Without Boyle's law no one would have known of the need to add the refinements which led to van der Waal's equations, and thus to a more adequate treatment of the behaviour of gases at higher temperature and pressures.

Of course, if the exceptions are enormous one might well query the value of one's original hypothesis – although, how enormous this must be before it is significant, is a good point. Fortunately, this is hardly a worry in the case of the sociobiological explanation of human sibling incest barriers. Most people are not ancient Egyptians or Hawaiians, and most people are not royalty or nobility. And, of those that are, most keep their hands off their siblings. If I were a biologist or a physicist I would be most grateful for a claim which had as few exceptions as the sociobiological incest thesis.

Let me take up one final objection. Members of some societies are simply disinclined to enter into incestuous practices. Members of other societies back up the disinclination with explicit prohibitions – taboos – that are frequently enforced with severe laws. Is not the sociobiological position crippled by its inability to speak to taboos, staying merely with disinclinations? Is not the sociobiological position ruined by taboos, in that their existence makes disinclinations (and hence biology) unnecessary?

I confess that I do not see that the existence of taboos forces upon us a non-inclusive disjunction. Given that humans have linguistic and related abilities – most unlike the brutes – and given that these abilities also have their adaptive roles, I see no reason whatsoever to suppose that, what first evolved non-linguistically, cannot be backed by the forces of our later-evolved speech and thought powers. Writing love poems hardly takes the sex out of romantic liaisons.

This is not to say that the sociobiologist can simply ignore taboos. They obviously need some explanations. It is to say that the existence of taboos is not in fundamental conflict with the basic position. As it happens, at least one biologist has gone on to consider taboos. Patrick Bateson hypothesizes a two-stage process of evolution: first, the barriers, then the supporting prohibitions.

He writes:

The historical argument can be summarised as follows.

1. Humans evolved like many other animals so that experience obtained in early life provided standards for choosing mates when they were adult. Normally the experience was with close kin so that, when freely choosing a mate, a person preferred somebody who was a bit different but not too different from close relations. This system evolved because those who did it avoided the maladaptive

costs of both extreme inbreeding and too much outbreeding. Consequently, the system was more likely to be represented in subsequent generations. This step could have occurred long before language evolved.

2. For other reasons to do with the benefits of cohesion, conformity was enforced on all members of the social group. Those who did things that the majority would not do themselves were actively discouraged. In sexual affairs, pressure was put on people who interested themselves either in individuals who were very closely related or who were strange to members of the group. As language evolved, prescriptions about mating were transmitted verbally to the next generation. In this way taboos and marriage rules characteristic of a culture came into existence. (Bateson 1986b, pp. 13-14)

This is just a hypothesis, and a number of questions are left blank. For instance, what is the nature and cause of the supposed group conformity? Presumably, we have to suppose that, grounded somewhere in their biology, humans have dispositions which incline them (or at least make them susceptible to) group norms. We may not have sheep genes, but we have to have herd-like instincts of some kind. (This is *not* to say that we are all blind conformists.)

Enough has been said. Strengths and weaknesses of the sociobiological position on incest have now been laid out. I suggest that it is reasonable to conclude that, essentially, the sociobiological position on incest is true and enlightening. Since, as noted, the treatment of incest is central to the sociobiological approach, this constitutes a strong reason for regarding the general approach with favour.

THE KIPSIGIS

Yet, even if you agree with me about the particular question of sociobiology and incest, you may still be feeling some unease. There is, perhaps, something a little *post hoc* about the whole discussion. Long before the coming of human sociobiology, we knew already that there are incest barriers. And even in the nineteenth century someone had linked them to biology. Thus, however strong the case may be, we are working on old problems, from behind. If human sociobiology is really to capture our imagination and support, let it show that it can push us into new areas. Even if only sociologically, let it show that it has the fire in its belly of a new paradigm. Let it show – and this, for me, is the crucial test – that it can attract the attention of bright, ambitious, graduate students.

In fact, such a demand is not quite fair on the incest case. There is more to human blood relationships than brothers and sisters, or even than parents and children. Sociobiologists are trying to fit incest into a wider, comprehensive picture, which does go well beyond the already-known or long-suspected. As is hinted in the quotation from Bateson just above, there is (for instance) much interest in the virtues of not going too far from the central family in the search for mates (Hartung 1985). But, to make the sociobiologist's case as strongly and distinctly as possible, I shall now switch, and shall look briefly at the ongoing work of one who is indeed at this very time of writing (August, 1986) still a graduate student. What I hope to identify is an exploration which is less sweeping than the incest explanation – quite fine-grained, in fact – but which shows that human sociobiology has enough appeal 'to attract an enduring group of adherents away from competing modes of scientific activity. Simultaneously, it [is] sufficiently open-ended to leave all sorts of problems for the redefined group of practitioners to resolve' (Kuhn 1962, p. 10).

Monique Borgerhoff Mulder focuses on the Kipsigis, pastoral people living in South-Western Kenya (Borgerhoff Mulder 1987a, 1987c, 1988) (see Figure 7.2). They raise cows and goats, grow corn (maize) for their own use and for sale, and the men (occasionally) hire themselves out as manual labourers. Land is held by individuals (men), and passed from father to son(s). The Kipsigis are polygynous, with some men having several wives and others none at all. To obtain a wife, bridewealth must be paid to a woman's family – on average, this payment consists of six cows, six goats, and 800 Kenyan shillings. In our terms, this translates into about £450 (US $700), and is a fairly substantial investment – more than a third of the average man's wealth.

179

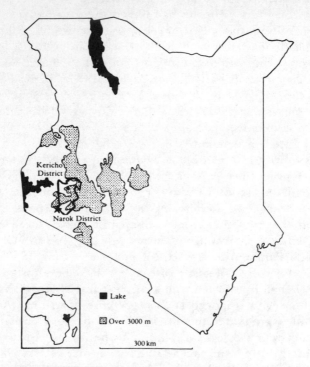

Figure 7.2 Location of the site of the Kipsigi study.
(Exact location marked with a star, major town with a dot, and
total district is included in box.)

It is usual for a father to finance a son for the first wife; but
if later (on average, ten years later) a son wants a second wife,
he must pay for her himself. Men first marry around the age of
23 (that is, those from a family rich enough to pay bridewealth).
Girls marry around the age of 15; this is always after, but not
long after, their circumcision, an event which follows menarche.
Decisions about marriage are economic, and have nothing to do
with Western notions about romantic love and the like. Divorce
and adultery are virtually unknown.

Borgerhoff Mulder learnt the Kipsigis language and lived with
them for eighteen months (June 1982-December 1983). She took
full census data over a study site of some 25 sq. kms, an area
covering several neighbourhoods. Using sociobiological theory,

she proposed and tested a number of hypotheses, of which I shall pick out two for discussion here.

First and most obviously, she raised a cluster of questions about people's aims and successes in society, and the reproductive payoff of such aims and successes. If sociobiology is to have any bite at all, then what people go after should pay off in terms of offspring – although, of course, biologically speaking it is not necessary that people recognize fully what they are about. Borgerhoff Mulder therefore considered the hypothesis of whether Kipsigis (or, more particularly, Kipsigis men, since they have the more overt and easily studiable behaviour) strive for and if successful achieve, goals that lead to more offspring. And if they do, is there a causal connection and what is it?

More formally, Borgerhoff Mulder tested a hypothesis of her supervisor William Irons (1979), that 'striving for conscious cultural goals is a proximate means of attaining unconscious reproductive ends'. This is a special instance of a more general theoretical principle that 'different patterns of culture and social organization are the evolutionary outcome of the propensity of humans to modify their behaviour in such a way as to promote most effectively their reproductive interests in any given environment' (Irons 1979).

There was little difficulty in determining what Kipsigi men wanted – they wanted what the rest of us want: wealth, power, influence, and the like. This could be determined by personal observation, questioning, and a host of other means. Moreover, Borgerhoff Mulder found that status could be accurately determined by just one measure, namely the number of acres a man held: peoples' perceptions of themselves and of others correlated tightly with this measure. Not surprisingly, on average, older men tended to be better off than younger men; but, since Kipsigi men belonged to cohorts (based on their year of circumcision), she was able to control for this factor by considering relative success within cohorts. Land holdings were the key to assessing aims and successes.

But what is the pay-off in biological terms? Quite simply, men who are successful in their cultural goals tend to be the men who have the most children. More precisely, they are the men who have most children who survive through to reproductive maturity. (By considering older cohorts, Borgerhoff Mulder was able to disregard the high infant and child mortality.) 'Significant positive correlations are found in all cohorts between wealth of a man (as measured in acres) and both the number of his surviving offspring and the rate at which surviving offspring are born per married year' (1987c, p. 12). Underlining this point, I reproduce, figuratively, the data from one cohort, the *Maina* (consisting of forty men who were circumcised between 1922 and 1930) where several men have twenty or more children, and one has over eighty! (Figure 7.3).

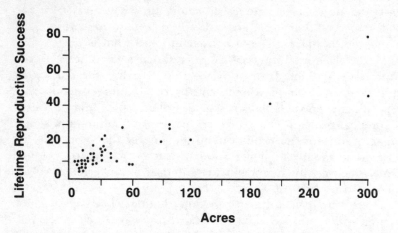

Figure 7.3 Comparison of male reproductive success with land-holdings.

Wherein lies the causal connection? If sociobiology is true, one would expect to find that cultural success leads to reproductive success. But (assuming that such consistent correlations are not pure chance), perhaps the causal connection goes the other way, or the correlation is due to some third, hitherto unmentioned factor. Perhaps, having lots of children enables a man to acquire new land, or perhaps land and children are separately due to educational achievements. However, Borgerhoff Mulder was able to exclude these and like rival suggestions, leaving strong the claim that cultural success leads to biological success. For instance, she could show that younger, culturally successful men, whose own families were not yet able to help, still did better biologically. And factors like education seemed to have no effect on biological success.

In terms of the discussion in the early part of this paper, what is really needed to forge the link between biology and behaviour is some filling in of the proximate causes. Why does the possession of wealth, particularly land, lead to more offspring? Given Kipsigi polygyny, the answer is simple and readily confirmed. Men with more land can afford more bridewealth payments, can acquire more wives, can father more children. It is possible to follow the chain through with relative ease, and thanks to the almost Victorian sexual standards of the Kipsigi, other causes can be discounted or disregarded. For instance, there is little chance for culturally unsuccessful males surreptitiously to engage in sexual activity.

Let us turn now to a second hypothesis proposed and tested by Borgerhoff Mulder. What factor or factors controlled the size of the all-important bridewealth? Again, there was a fairly straightforward sociobiological prediction: A man (or his father) should pay more for a girl who will give him more children or (indirectly) otherwise help his reproductive abilities. Given the importance of property in acquiring wives (for a man himself or for his sons), and given that it was the Kipsigi women who were the family work-horses, a woman who was potentially a hard worker had obvious sociobiological virtues.

Guided by such considerations as these, Borgerhoff Mulder found that there were two factors which enabled her to express her hypothesis more precisely. First, there was a firm correlation between the age at which a girl was circumcised and her life-time reproductive success (see Figure 7.4.). A girl who came to menarche earlier was circumcised earlier. She thus got married earlier and had more children. Second, girls who lived close to their mothers spent more time at mothers', and hence were less productive around their husbands' homes.

The sociobiological hypothesis can now be expressed quite strongly. Men (or their fathers) should be expected to pay more for younger brides and for those who live further from the home of their birth. And this prediction is found to hold strongly: 'Bridewealth decreases with age ... with the main difference lying between those brides who marry before and after reaching 16 years old' (Borgerhoff Mulder 1988, p. 6) (see Figure 7.5). Likewise: 'There is a progressive increase in the bridewealth paid between adjacent houses, between families in the same

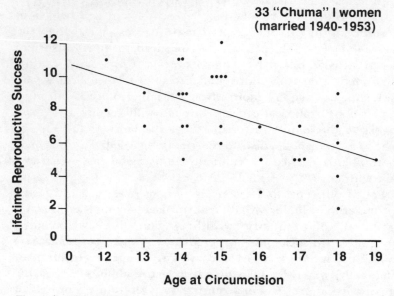

Figure 7.4 Comparison of female reproductive success with age of circumcision.

kokwet [neighbourhood], between families outside the *kokwet* but within 8 kilometre range'. (See Figure 7.6.)

As before, Borgerhoff Mulder was able to exclude other hypotheses. For instance, no more is paid for an educated bride than for an uneducated one. Moreover, again one can start to forge the proximate links. Men are acutely aware of – and frequently much annoyed at – the time wives spend with their mothers. Apparently, they are less sensitive to their brides' ages, let alone their reproductive potential. But they do favour fat women over skinny ones, and this can be linked both to age of menarche and to nutritional and general health status.

For our purposes, there is no need to go into further detail. Borgerhoff Mulder's work has restricted scope. She makes no sweeping statements, across cultures or across time. Furthermore, her theoretical underpinning is not made fully explicit (although see Borgerhoff Mulder 1987b, Caro and Borgerhoff Mulder 1987). Her position seems to be that humans are organisms that can respond to environmental (including

161 marriages (post 1959)

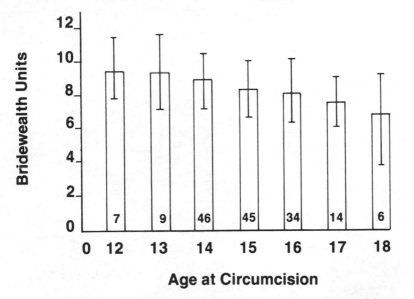

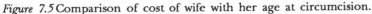

Figure 7.5 Comparison of cost of wife with her age at circumcision.

cultural) change, maximizing reproductive options, without the need for wholesale genetic change. This, presumably, is because our biology, specifically that leading to and including our thought processes, has a built-in flexibility – selection has worked (and still works) to make us agile, rather than fixed and determined by our genes. However, Borgerhoff Mulder really provides no proof of her assumptions, other than to note that now (for various reasons) polygyny is starting to become more difficult, successful men are starting to pay yet more attention to the quality of potential wives. One supposes, given human reproductive practices, this flexibility is limited – but where and when the adaptive turns to the maladaptive is not specified. (Nor could it be, on such data as Borgerhoff Mulder has.)

But, against all of this, one must state firmly that Borgerhoff Mulder makes the case she sets out to make. Moreover, she does it with great methodological sensitivity (almost over-sensitivity). You cannot fire easy shots, claiming that she has ignored or played down obvious confounding factors. She shows that,

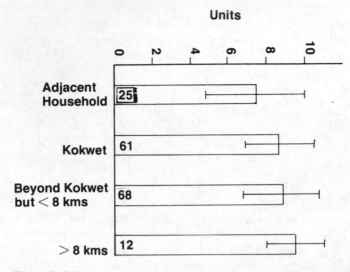

Figure 7.6 Comparison of cost of wife with husband's distance from her parental home.

against a sociobiological background, important things done by the Kipsigis make good sense. This, I would suggest, is no mean praise for human sociobiology (or human sociobiologists) at this stage of its development.

Indeed, Borgerhoff Mulder's work is impressive; I am not sure how much more one could ask of sociobiology and its students. Work like this shows that human sociobiology has a rightful place in the evolutionary family.

CONCLUSION

Were I simply just offering a survey, I would provide ten or twenty more items of work like Borgerhoff Mulder's. A rapidly growing number of people are studying human behaviour, particularly human social behaviour, relating it to our biology. There is a range of projects, running the gamut from the activities of the younger sons of late-medieval Portuguese noble families to the child-beating propensities of step fathers in

Hamilton, Ontario. (A good review is Betzig *et al* 1988.) Some of the most interesting work – certainly some of the most controversial – has centred on attempts to provide formal models of the ways in which genes and culture can interact and evolve over time. (See, for instance, Cavalli-Sforza and Feldman 1981, Lumsden and Wilson 1981, 1983, Boyd and Richerson 1985.) Detailed consideration would be too much of a side-track here. In their present form, I doubt whether the models will stand the test of time – but then, how many such models (in any area of science) do? Their virtue probably lies in their attempt to beat a path, which can then be improved by others. Human sociobiology has had a controversial decade. The final verdict is not yet in. The prospects are bright and grow ever brighter. And, already, we are reaping the first harvests of this revolutionary way of understanding ourselves and our relations to the world around us. But, ultimately, words in defence of human sociobiology are ineffectual or unnecessary. Hence, at this point I will cease my advocacy and turn to other discussions, where I can further the work at hand.

NOTES

1 People mean lots of different things by 'determinism', especially when used as a term of abuse against human sociobiologists. I try to tease out some of these meanings in Ruse, 1987b.
2 This is because animals have the same genetic mechanism as we have. Inbreeding problems arise because hitherto masked genes can make their effects known: technically, 'recessive' genes are paired 'homozygously'. Unfortunately, many recessive genes are defective and they simply do not code for required cell products. So long as they are paired with functioning 'dominant' genes, no problems arise. But when they are on their own trouble starts.

Chapter Eight

BIOLOGICAL SCIENCE AND FEMINIST VALUES

My topic is the relationship between science and values. Specifically, I will look at biological science, especially evolutionary theorizing, and at those values held dear (not to mention those values abhorred) by supporters of the feminist movement. Since discussions of this nature rapidly take on all of the fervour of a revivalist camp meeting, it is appropriate to begin with a text. I draw mine from *Valley of the Amazons*, the recent novel by Noretta Koertge. The conversation is between two women: Helen who is militantly anti-male, and Tretona, the heroine, who keeps one foot in the real world (of a kind), namely a philosophy of science department.

> 'So is the claim that men evolved after women? That's pretty speculative, isn't it? Just sounds like an inversion of the old Adam-before-Eve story to me.'
>
> Helen's eyes flashed but her answer rolled out smooth as silk: 'Well, you obviously haven't read [Elizabeth Gould] Davis' book. All evolutionary scenarios are speculative, of course, but Davis makes out a pretty good case. The Y-chromosome is just a broken-off bit of the X, you know. And of course almost all academic biology has been done by men – although even they admit womyn were the first to domesticate plants and animals.'
>
> 'But having two sexes is a very common evolutionary strategy. It's a good way of getting genetic diversity. Gee, you find sexual dimorphism in all sorts of lower animals and plants. Sex is a lot older than people. You can't just say the

Y-chromosome is a little castrated mutation of the X. That doesn't make any biological sense at all.'

'Well, it makes a lot of sense to womyn.' And with considerable feeling Helen told about how liberating these ideas had been to the Ovular. 'We had already seen how early matriarchal forms of social arrangements had been smashed by the violence of war and patriarchal religions and then to see the parallel in biological development – it was all very impressive.'

'But it's phony biology. You're just spinning myths.'

Now, Helen was really serious. 'Don't ever doubt the power of myths, Tretona – I mean womyn's myths, not the bullshit stories we grew up with. We have to build a feminist mythology and recover womyn's herstory. What difference does it really make how the two are intertwined? Both are sources for spiritual growth and political action.'
(Koertge 1984, pp. 37-8)

I suspect that most conventional philosophers of science will recoil from this passage with horror. It will be felt, as Tretona (who has good Popperian leanings) feels, that you simply cannot spin things around as Helen suggests, and stay with anything remotely resembling science. Real science is about objective, disinterested, 'external' reality. It is about the way that the world *is*, rather than about the way that the world *ought* to be. To quote Ernest Nagel:

There is a relatively clear distinction between factual and value judgments, and ... however difficult it may sometimes be to decide whether a given statement has a purely factual content, it is in principle possible to do so .... [I]t is possible to distinguish between, on the one hand, contributions to theoretical understanding (whose factual validity presumably does not depend on the social ideal to which a social scientist may subscribe), and on the other hand contributions to the dissemination or realization of some social ideal (which may not be accepted by all social scientists). (Nagel 1961, pp. 488-9)

Recently, however, a growing number of writers on science have started to question this happy, safe line that Nagel draws between science and values. Particularly, feminist writers ask whether there might not be more to Helen's position than conventional philosophers would allow. That which passes for science is so value-impregnated – usually, so male-chauvinist value-impregnated – that any pretence at objectivity is merely hypocritical self-justification. 'Science', so called, is little different from that which we label 'myth'.[1] (In Freud's case, not at all different.)

In the hope of pulling out what I take to be of genuine worth in the feminist critique, and yet in the hope of preserving some of what the conventional philosopher of science (like Tretona and Nagel) sees as distinctive and important about science, I want first to look at two recent feminist discussions of science. These discussions deal specifically with aspects of evolutionary biology – a nice subject for discussion, standing as it does between the ethereal heights of the physical sciences and the murky depths of the social sciences. Then, I shall set these discussions in context, using here as guide-lines the most stimulating 1982 Presidential Address to the Philosophy of Science Association, by Ernan McMullin. I shall explore my findings, looking at their value implications for actual (evolutionary biological) science.

FEMINIST CRITIQUES

I begin with a highly critical review of evolutionary theorizing, past and present, by the biologist Ruth Hubbard (1983). She argues, illustrating her case with copious quotations, that the Darwinian theory of evolution through natural selection has, from the time of its birth until the present, been little more than a vehicle for male-dominant ideological sentiments, where females are seen merely as passive playthings or helpmates for the all-important active penis-possessing members of the tribe.

Start with Darwin. The most explicit example of his blatant sexism lay in his secondary evolutionary mechanism, so-called 'sexual selection' (Darwin 1859, 1871). This involves competition within the species, for mates of the opposite sex. 'The Victorian and androcentric biases are obvious', writes Hubbard, going on to make her case with quotations like the following:

> Generally, the most vigorous males, those which are best
> fitted for their places in nature, will leave most progeny.
> But in many cases, victory depends not so much on
> general vigor, as on having special weapons confined to
> the male sex. (Darwin 1859, p. 88)

And

> With animals which have their sexes separated, the males
> necessarily differ from the females in their organs of
> reproduction; and these are the primary sexual characters.
> But the sexes differ in what Hunter has called secondary
> sexual characters, which are not directly connected with
> the act of reproduction; for instance, the male possesses
> certain organs of sense or locomotion, of which the female
> is quite destitute, or has them more highly-developed, in
> order that he may readily find or reach her; or again the
> male has special organs of prehension for holding her
> securely. (Darwin 1871, 1, p. 245)

Consequently, it is no surprise that:

> Man is more courageous, pugnacious and energetic than
> woman, and has more inventive genius. (Darwin 1871, 2, p.
> 316)

Hence, it follows that:

> The chief distinction in the intellectual powers of the two
> sexes is shown by man's attaining to a higher eminence, in
> whatever he takes up, than can women – whether requiring
> deep thought, reason, or imagination, or merely the use of
> the senses and hands. If two lists were made of the most
> eminent men and women in poetry, painting, sculpture,
> music (inclusive both of composition and performance),
> history, science, and philosophy, with half-a-dozen names
> under each subject, the two lists would not bear comparison.
> We may also infer ... that if men are capable of a decided
> pre-eminence over women in many subjects, the average of
> mental power in man must be above that of woman .... [Men

have had] to defend their females, as well as their young,
from enemies of all kinds, and to hunt for their joint
subsistence. But to avoid enemies or to attack them with
success, to capture wild animals, and to fashion weapons,
requires the aid of the higher mental faculties, namely,
observation, reason, invention, or imagination. These
various faculties will thus have been continually put to the
test and selected during manhood. (Darwin 1871, 2, p. 312)

One's immediate reaction is to dismiss all of this as a sad legacy
of the nineteenth century. But Hubbard then proceeds to argue
that neo-Darwinism, in its late-twentieth-century version,
incorporates much the same set of values. Most particularly, that
bête noire of so many feminists, sociobiology, comes under
Hubbard's stern gaze. It is argued by modern Darwinian
biologists (in a tradition going directly back to Darwin himself),
that natural selection plays a significant causal role in animal
social behaviour, and that in this respect our own species must
be firmly included. The ways in which we think and behave owe
much to our evolutionary heritage, as manifested through our
underlying genetic framework. We do what we do and we think
what we think, in major part because so doing and thinking
proved reproductively advantageous to our ancestors (Wilson
1975a).

Again Hubbard finds rampant sexism. Thus, Wolfgang Wickler
on lower animals:

Even among very simple organisms such as algae, which
have threadlike rows of cells one behind the other, one
can observe that during copulation the cells of one thread
act as males with regard to the cells of a second thread,
but as females with regard to the cells of a third thread.
The mark of male behavior is that the cell actively crawls
or swims over to the other; the female cell remains passive.
(Wickler 1973, p. 23)

George Williams on higher animals:

Males of the more familiar higher animals take less of an
interest in the young. In courtship they take a more active
role, are less discriminating in choice of mates, more

inclined toward promiscuity and polygamy, and more contentious among themselves. (Williams 1975, p. 1)

Edward O. Wilson on all animals:

One gamete, the egg, is relatively very large and sessile; the other, the sperm, is small and motile .... The egg possesses the yolk required to launch the embryo into an advanced state of development. Because it represents a considerable energetic investment on the part of the mother the embryo is often sequestered and protected, and sometimes its care is extended into the postnatal period. This is the reason why parental care is normally provided by the female. (Wilson 1975a, pp. 316-17)

About this last passage, Hubbard writes as follows:

Though these descriptions fit only some of the animal species that reproduce sexually, and are rapidly ceasing to fit human domestic arrangements in many portions of the globe, they do fit the patriarchal model of the household. Clearly, androcentric biology is busy as ever trying to provide biological 'reasons' for a particular set of human social arrangments. (Hubbard 1983, p. 60)

There is more; but the thrust of Hubbard's argument is clear. The reader is left in little doubt why, in her title, she asked sardonically: 'Have only men evolved?'

Thanks to the values she finds, Hubbard's tale is sombre and pessimistic. Complementing her discussion, I turn next to a writer who finds no fewer values in evolutionary biology, but who nevertheless sees cause for hope. The writer is the sociologist Donna J. Haraway, and her tale is one of women active today in the field of primatology. Although Haraway's approach is very much that of the social scientist, her overall interest is that of the philosopher. 'In particular, does the practice of their science by women in a field of modern biology-anthropology substantially structure discourse in ways intriguing to feminists? Should we expect anything different from women rather than from men?' (Haraway 1983, p. 176).

In response to this question, Haraway tells a fascinating story. The key figure in the development of modern primatology (in

America) was Sherwood Washburn. In the 1940s, he revolutionized his subject, taking it from its then anatomical emphasis (an ideal vehicle for all kinds of unsavoury theorizings about the differences between the races) to a new home within the evolutionary family, where there would be lots of emphasis on field work, and integration with the rest of modern biology.

> By the mid-1940s Washburn was practicing physical anthropology as an experimental science; by 1950 he was developing a powerful program for reinterpreting the basic concepts and methods of his field in harmony with the recent population genetics, systematics, and paleontology of Theodosius Dobzhansky, Ernst Mayr, and George Gaylord Simpson. (Haraway 1983, p. 182)

Washburn was successful not only in his theoretical aims (for example, Washburn 1951), but also in building around himself a band of committed student disciples. These in turn have had students, and so today we find primatology stocked, influenced, and outrightly controlled by Washburn's intellectual children and grandchildren. And included in this still-growing family are many women. It is Haraway's thesis that in the work of these women we see reflected many of the moral, political, and ideological struggles that have engaged (middle-class, intellectual, white, American) women in the past several decades. All of the scientific stories Haraway examines were 'shaped materially by contemporary political struggles, in particular conflict over the reproductive social behavior of women in the last quarter century' (p. 201).

Haraway sees the original (Washburn) position on primate social life (with obvious implications for human evolution and our present state) as one focusing on male dominance. Supposedly, male primates (Washburn's own subjects were baboons) strive for position within the troop, gaining access to and controlling the females. All surrounds and follows from this male activity. Specifically, in the human case, through the vision of 'man the hunter', we have the causal clues to our distinctively peculiar evolution, as we became bipedal and greatly increased our intelligence. Those proto-human males with (what today we think of as) peculiarly human features were precisely those males who succeeded best in life's struggles – as they hunted for

194

themselves and their families, thus supporting and guiding the proto-human womenfolk.

Then, Haraway tells how, from the 1960s on, we get the entrance and growing influence of the women. And, with this, slowly but definitely we get a change in the scenario. Gradually the role of female primates gains more attention, and corresponding importance. Thus, for instance, Sarah Blaffer Hrdy (a student of a student of Washburn) follows the male-dominant pattern in seeing males (her subjects were langurs) moving in on females, killing the infants that these females have mothered. In this, Hrdy shows not only the influence of (by now) conventional primatology, but also the influence of sociobiology, for she believes that the male infanticidal practices further the males' own reproductive ends. The females, deprived of their infants, more readily come back into heat, thus opening up reproductive opportunities for the newly successful males.

However, it is crucial to Hrdy's account that the females also have vital social roles. They too have reproductive interests. Thus, they actively strive to prevent infanticide. 'Infanticide in certain circumstances became a rational reproductive strategy of langur males, opposed to a rational extent by langur females, whose reproductive interests were certainly not the same as the males' (Hrdy 1977, p. 192). Hence, at least we have a female response in primate social behaviour – a female response which is no less significant than the male behaviour. 'Indeed, sexual conflict from a sociobiological viewpoint is a necessary consequence of sexual reproduction' (1977, p. 193).

Concluding her account, Haraway tells us that, recently, there are female primatologists who have gone much further even than Hrdy. These researchers quite eliminate the Washburnian emphasis on the all-important role played by males. Now, if anything, it is the females who rule the roost – literally and metaphorically. Male 'takeover' of troops is seen as 'rapid social change'. And male infanticide is seen, less as an adaptive male reproductive strategy, and more as a pathological instability – of help to no one.

Haraway quotes the primatologist Jane Bogess:

In certain populations, where there is social crowding and artificially high densities and where adult males live outside

bisexual troops, the species-typical characteristic male social instability can operate against the reproductive success of all troop members, including the new male residents.
(Bogess 1979, p. 104)

We have come full circle. Although Bogess is no less a Darwinian than Washburn and Hrdy, we have the makings of a herstory of human evolution that would satisfy even Koertge's fictional Helen. Haraway herself makes no explicit evaluations; but she does admit to having favourite primate stories, and we are left in little doubt that, for once, newer is taken as better.

CONVENTIONAL RESPONSES

So what are we to make of these two perspectives on Darwinian evolutionary theory? Three obvious conventional responses can be dismissed.

First, it will be said that the facts (i.e., the elements of science) really cannot be as Hubbard and Haraway describe them. It will be said that there must have been grossly selective and distorting treatments of the appropriate materials. Darwin was a great innovative scientist, and so he cannot have been as crude a chauvinist theorist as Hubbard maintains. After all, Darwin was a revolutionary, no less than (and probably rather more than) most contemporary feminist heroines.

A similar defence will be offered for modern biology. It will be seen as ridiculous to suggest that someone talking about algae is a sexist – at least, that someone talking about algae thereby shows their sexism. And the same will be said of Haraway's imaginative reconstruction of recent primatological history. No doubt one or two people may have let their enthusiasms run away, but to pretend that ideology has been as influential as Haraway suggests is just plain ridiculous.

Now, responding to the response, there is certainly some truth in the conventional rebuttal. Hubbard's treatment of sexual selection is less than fair. She concentrates exclusively on sexual selection through male combat, omitting entirely mention of sexual selection through female choice. Darwin (1859, 1871) argues that much sexual dimorphism – for instance, the tail possessed only by the peacock and not the peahen – is caused by the female choosing the mate she prefers. Here, it is the male who is passive and the female who is active – hardly the universally male-dominated picture of Darwinism that Hubbard paints.

Again, an unfriendly critic might point to the fact that although Hubbard finds sociobiology tainted with sexism, Haraway notes that in major part it was sociobiology that enabled Hrdy to break from Washburn's male-dominated picture of primate family life! The conventional philosopher of science can therefore properly complain that you cannot have things both ways – at least, you should not. Either sociobiology is at fault, or it is not.

However, these are quibbles. If you read Darwin, again and again you are struck by how much and how firmly he was a child of his time. Indeed, as scholars now recognize, Darwin succeeded because he was an archetypal Victorian – not because he was not (Ruse 1979a, Ospovat 1981). Darwin's major work on our own species, *The Descent of Man*, is an unbroken hymn to male dominance and success – except, that is, when it is belittling the blacks, or running down the Irish, or extolling the virtues of capitalism (Ruse, 1985). Hubbard's quotations can be matched, time and again.

Modern evolutionary biology is a lot less blatant in its authors' biases than are the writings of Darwin and his contemporaries. However, even here Hubbard is right in her general claims that such biases do often show through. I am not sure that her own examples are always the best, but many can be found. My own favourites are the two reproductive strategies highlighted and labelled by Richard Dawkins in his *The Selfish Gene* (1976). Apparently, if animals do not opt for the 'He-man' strategy, they are liable to be found taking the 'domestic-bliss' strategy. If this is not to reveal certain attitudes towards male/female statuses and respective worths, I do not know what is.

Finally, even a casual reading of recent articles and books by today's primatologists fully justifies Haraway's interpretation of changed emphases, reflecting altered value assessments and commitments. Moreover, (partially in defence of my own sex) it is worth pointing out that not only female primatologists now make much of the role and significance of the female apes in the day-to-day life of primate packs. For instance, the Dutch ethologist Franz de Waal has recently reported on his team's studies of a band of (some fifty) semi-wild chimpanzees at the Arnhem Zoo. The group lives in a state of constant tension, as the males strive for pack dominance. But the females are no less active and important. Male success is absolutely dependent on the support of the females – even the beginnings of male conflict start with the females. Moreover, group stability is almost entirely a function of the harmonizing abilities of the females, particularly the older ones. No one reading de Waal's story, *Chimpanzee Politics* (1982), has any doubt as to the key importance of the females of the group.

The values are there – which, at once, brings up the second conventional response. Obviously it will be said that what we are dealing with is bad science or non-science. Good science – genuine science – contains no such blatant impregnation by moral or ideological values, whether these be of the kind that Hubbard abhors or of the kind that Haraway endorses. Science is about the objective world; about the real differences between males and females; not about whether you approve of these differences or think they support your convictions that males or females are of more inherent worth than members of the other sex.

The trouble with this response is that it wipes out a great deal of what we commonly call 'science' – work that is produced in university 'science' faculties, that passes as and is looked to as 'science', and (as Haraway shows well) that serves as a role model for would-be 'science' students. You may not like Darwin's views, you may like (say) Hrdy's views, but the all-important point is that these people produce what we know as 'science'. To argue otherwise suggests that you are simply defending your philosophical thesis by fiat.

Moreover, the suspicion is that your model for perfect science is physics – where, at least, values of the kind we are talking of are a lot less obvious. In defence of the feminist critique: the time has surely come to recognize that biology is a science in its own right, and that this being so, the values – good, bad, and indifferent – are right there in it.

The third conventional response, related to the second, is to admit the values in past and present evolutionary biology, but to deny that they are inevitable or that they are a good thing – whether you yourself approve or disapprove of the values. This response argues: in its way, Hrdy's work is just as flawed or immature as was Darwin's.

But, apart from the fact that this is an exceedingly condescending response – who are we, as philosophers, flatly to label much of biological science as 'immature'? – this response fails for much the same reason that the second response failed. Unless some reason can be given for the immaturity of value-impregnated biology, the case is being made by little more than a priori fiat. A model for science – probably physics inspired – has been set up and then conformity to it insisted upon. Unless independent reasons are given, the case does not convince.

The values in evolutionary biology are not to be dismissed so readily. What then are we to do and say? Must we swing right to the other pole, agreeing with Helen that science is no more than myth anyway, and that basically anything goes? Could it be that values are at least as important, if not more important, than anything else? Get your ideology right, and no other thing really matters.

This option seems no less happy than that of the traditional philosopher of science. After all, virtually by definition, science, including evolutionary biology, does seem to be about the external world. Science is not just a matter of feelings and emotions and values, important as these undoubtedly are. The very aim of science is to tell us something of reality. Melding science and myth misses the point of the whole scientific enterprise. At least, even if science does collapse into myth, we need more argument to show that all of scientists' good intentions necessarily come to naught.

Certainly, for all of their critical comments, this 'realism' is the position endorsed by the feminist writers on science. Thus Hubbard dismisses the efforts of those women who would spin myths à la Helen, simply making the female of our species far more significant than the male. It is true that this dismissal comes, in part, because Hubbard believes authors like Elaine Morgan (*The Descent of Woman* 1973) and Evelyn Reed (*Woman's Evolution* 1975) implicitly carry on many of the offensive male values. But, Hubbard feels also that science is different, somehow transcending values, and being controlled by the real world. 'Women can sift carefully the few available facts by paring away

the mythology and getting as close to the raw data as possible'
(Hubbard 1983, p. 66).

Similarly, Haraway thinks of science as more than mere
storytelling, done according to your ideological perspective or
political value system.

> To ignore, to fail to engage in the social process of
> making science, and to attend only to use and abuse of
> the results of scientific work is irresponsible. I believe it is
> even less responsible in present historical conditions to
> pursue antiscientific tales about nature that idealize
> women, nurturing, or some other entity argued to be free
> of male war-tainted pollution. (Haraway 1983, p. 200)

Assuming, therefore, that science really is (if only in intention)
different from other human activities/products, and assuming
that this difference does lie in its attempt to grapple with and
understand the world of experience, we must find some place
for the putative objective, real-world mirroring elements in
science. At the same time, we must not compromise the place
of values. And more than mere values – we must find place for
political and social and moral values, of the kind we have been
discussing thus far. To aid our enquiry, I turn now to the
thoughts of Ernan McMullin.

MCMULLIN ON VALUES IN SCIENCE

McMullin (1983) begins with the crucial premise that empirical
facts (meaning by these no more than the experiences we have
of the world about us) underpin our scientific theories or
hypotheses. In science, you are always going beyond what you
directly perceive or know – peering into the past, the future, the
smaller, the greater. This means that, necessarily, values of some
sort enter in, as you try to bridge the gap between experience
and theory. And these values are not just about the morality of
your scientific habits (should you report the results of your
graduate student as if they were your own?) but values which,
in some sense, must enter into the moulding and constraining
of your finished product.

What is the nature of these values? McMullin picks out
several broad categories, of which I will highlight three, ordering

and treating them in ways which will aid our own enquiry. First, we have *pragmatic* values. Here, we are looking at values emerging from the kinds of external constraints that a scientist may have laid on him or her, by funding or time or the like. You have to make practical value decisions, and these will affect your finished science. Second, we have what McMullin calls *epistemic* values. Here, reference is being made to the general methodological rules by which the scientist agrees to be bound, and which he or she tries to follow. Among values of this kind, cherished by the scientist, McMullin lists: predictive accuracy, the hope to make forecasts which prove to be true; internal coherence, the aim to avoid contradictions; external consistency, namely the desideratum of not conflicting with other theories; unifying power, 'the ability to bring together hitherto disparate areas of inquiry' (McMullin 1983, p. 15); fertility, that is the power to make novel predictions and yield new insights; and simplicity.

Third, McMullin finds that a scientist is moulded by a veritable grab-bag of values, including those of particular interest to our feminist writers. These McMullin refers to under the bland general term of *non-epistemic* values.

> If values are needed in order to close the gap between undetermined theory and the evidence brought in its support, presumably all sorts of values can slip in: political, moral, social, religious. The list is as long as the list of possible human goals. (McMullin 1983, pp. 18-19)

How does McMullin fit together and assess all these various types of value in the scientific enterprise, and what implications does all of this have for our own particular enquiry? At the theoretical level, there is not much you can do about pragmatic values. These are very much a function of human weaknesses, not to mention the arbitrary dictates of grant-agency boards. But these are problems for all of science, and do not impinge in some distinctive way on our own field. This brings us on to epistemic values. It is here McMullin sees that which is central and important to the scientific enterprise. Indeed, it is not too much to say that, by definition, McMullin sees science as an activity which is governed by rules or criteria incorporating epistemic values. We attempt to come to grips with experience – the experience of our senses – and, in order to do so, we are guided by epistemic values. We aim for predictive accuracy, consistency, and the like.

The question now arises of how we distinguish the scientific enterprise from mere myth-making. Even before we get to non-epistemic values (which will be tackled in a moment), we have the problem of defending science against the charge that it is merely a product of human wishes, desires, and emotions. Given epistemic values, wherein lies the *objectivity* of science? McMullin makes two points in defence of the non-mythological nature of (good, genuine) science. First, the values turn out something which works.

> We have gradually learnt from [our] experience that human beings have the ability to create those constructs we call 'theories' which can provide a high degree of accuracy in predicting what will happen, as well as accounting for what has happened, in the world around us. It has been discovered, further, that these theories can embody other values too, such values as coherence and fertility, and that an insistence on these other values is likely to enhance the chances over the long run of the attainment of the first goal, that of empirical accuracy. (McMullin 1983, p. 21)

Second, speaking in a way more tentatively but in a way more significantly, McMullin defends the objectivity of science despite epistemic values – because of epistemic values – in terms of his belief about the reality of the external world.

> I think that there are good reasons to accept a cautious and carefully-restricted form of scientific realism, prior to posing the further question of the objective basis of the values we use in theory-appraisal .... The version of realism I have in mind would suggest that in many parts of science, like geology and cell-biology, we have good reason to believe that the models postulated by our current theories gives us a reliable, though still incomplete, insight into the structures of the physical world. (McMullin 1983, p. 22)

The point seems to be that we accept an external real world. *Qua* scientists, we try to map/explain/understand this world. And the best hope we seem to have to accomplish our scientific end comes through epistemic-value-impregnated rules – the justification for which use lies in their success. After all, if you do not appeal to historical vindication, what else do you appeal to?

There are obviously many questions which can and should be asked here. McMullin himself is highly sensitive to the thinness of the philosophical ice on which he skates. But I have nothing critical to say about McMullin's highlighting of the scientist's heavy commitment to epistemic values, or employing the defence of scientific objectivity despite/because of such values. This is partly because such a discussion would take us way too far astray, but more because I have elsewhere defended at length a view very similar to that of McMullin (Ruse 1985).

The only point I want to make here is that, if we accept McMullin's position, then we can satisfy at least some of the intuitions of our feminist writers. Science is not myth; it is about reality; and the epistemic values preserve its integrity as science. If Helen's view of human evolution is a function merely of non-epistemic values, and fails epistemically, then it is certainly not good science, and probably not genuine science at all. Science is about the real world, and if you ignore the only guides we have, then myth is all you have left. But, it is not all that you *need* have left.

NON-EPISTEMIC VALUES

What of the values which interest the feminists? We have seen that McMullin's analysis acknowledges the existence of these kinds of values in science. Moreover, although they are not McMullin's main focus, I see no reason to conclude from his account that they are unimportant. Such values permeate scientific achievements, as one like McMullin, sensitive to the history of science, fully appreciates.

Nevertheless, McMullin does in some sense think history supports the conclusion that the existence – certainly the importance – of such values in science is a sign of immaturity.

> To the extent that non-epistemic values and other non-epistemic factors have been instrumental in the original theory-decision (and sociologists of science have rendered a great service by revealing how much more pervasive these factors are than one might have expected), they are

gradually sifted by the continued application of the sort of value-judgement we have been describing here. The non-epistemic, by very definition, will not in the long run survive this process. The process is designed to limit the effects not only of fraud and carelessness, but also of ideology, understood in its pejorative sense as distortive intrusion into the slow process of shaping our thought to the world. (McMullin 1983, p. 23)

McMullin's is not the conventional response, discussed and dismissed above. McMullin is not simply saying: 'Physics has no need of non-epistemic values. Go thou biologist, and do without likewise.' Rather, his claim is that, starting from the fact that science is an attempt to understand the real world, we find the application of epistemic values gradually pushes out the need or place for non-epistemic values. And this is the case, whether we speak of physics, of biology, or of geology.

Consequently, I take it that McMullin would acknowledge just about everything that the feminists writers pick out. Nevertheless, he would deny (what I see as) the feminists' ultimate conclusion, namely that non-epistemic values are important to science, and here to stay – so we had better get them right. Bringing an old distinction up to date, McMullin would see non-epistemic values important in the context of *discovery*; but starting to go as we move towards the context of *justification.*

Can we find some middle way, using McMullin's analysis and insights, but saving some of the urgency about non-epistemic values felt by the feminists? On the one hand, I cannot deny that (speaking from history) McMullin does have a strong point. Let me support him by giving an example of my own. In 1830, Charles Lyell started publishing his *Principles of Geology.* In this work, he argued for his well-known 'uniformitarian' thesis, namely that the world is in a steady state with the only causal geological forces being of an intensity and type that we see about us today – rain, sleet, erosion, deposition, earthquakes, and the like. *Prima facie,* Lyell had no truck with religion, denying the need for miraculous interventions of a catastrophic nature (like Noah's flood). Yet, we now know that, in the *Principles,* Lyell was trying in major part to reflect his commitment to the Unitarian deistic God, whose greatness lies in non-intervention. He had foreseen and accounted for all when He put the world in motion (Ruse 1979a).

Today, the uniformitarianism of geology owes nothing to Lyell's deism. Epistemic criteria can establish all. There is no need to rely on a Unitarian world-picture in excluding miracles as the prime causal elevating factors behind the Alps. It is enough to rely on the rules of prediction, consistency, and so forth. Together, they have yielded the theory of continental drift through plate tectonics (Ruse 1981e).[2]

In cases like this, and one can give lots of good evolutionary biological cases – think, for instance, how our understanding of the fossil record has evolved away from a progressive interpretation speaking to God's goodness and glory (Bowler 1976) – I am sure that McMullin is correct in his relative assessment of epistemic/non-epistemic values. Yet, on the other hand, I sense that there is still much truth in the feminist position.

First, it should be noted that (despite family resemblances) we really have had to move beyond the strict old 'context of discovery/context of justification' dichotomy. You can no longer pretend that non-epistemic values work during that initial flash of inspiration, but that as soon as you get something on the table – certainly, as soon as you get something into the journals – you drop off such values, and work solely in the epistemic realm. The feminists have surely established (strongly and successfully) that so long as science is live and vibrant – so long as science is at the cutting edge of active thinkers – non-epistemic values will be important. You might almost say that, as such values go from science, interest declines. The product remaining is important – fundamental – but it is the science of the textbook or undergraduate examination question. For working scientists, non-epistemic values are crucial. So it is legitimate to demand that we get them right.

Second, even given McMullin's appeal to history, I still wonder if non-epistemic values are/can be/should be ever eliminated entirely. There are at least three points at which I see non-epistemic values entering into science – points which confirm the feminist concerns with the nature of science.

First, non-epistemic values govern the very choice of *topic*. Science is not just about any old aspect of reality. Pragmatic factors are important here, for you cannot look at everything –

if indeed it makes sense to talk of studying all of reality. But, in making our decisions about what to study, as the feminists argue, we show non-epistemic influences. If you look only at males (or primarily so), you might satisfy all that McMullin demands, and yet clearly you show certain priorities. Thus, here, there is a continued need to urge proper moral sentiments.[3]

Second, non-epistemic values govern the choice of *language*, especially the metaphors which are used. You might do all that McMullin asks, and have all the epistemic success he desires, and yet show bias through your language. Suppose, for instance, you referred always to females as 'girls' and to males as 'men'. Again your priorities and sense of worth would show through. As I have agreed, this is no hypothetical concern here of feminists. Speaking of a 'domestic-bliss' strategy is demeaning.

You can never eliminate language from science. Perhaps, however, you might feel that metaphors die, in the sense that they lose their old force, and become defined purely technically and epistemically within the theory. Who now thinks of the physicists' 'force' as having anything to do with human or divine will-power or actions? I would concede that a draining out of the original meanings does undoubtedly occur. But, I see no reason why such must (or does) inevitably come to pass. Moreover, we should never underestimate the power of language to spring again to life. Twenty-five years ago, who would have dreamed of the extent to which terms like 'man' would become sexist symbols?[4]

Third, despite McMullin's claims, I suspect the very *models* or hypotheses of science can – in ways sensed by the feminists – continue to reflect non-epistemic values. For instance, as many have noted, including the feminist writers, much of evolutionary theorizing is based on models of competition, power, aggression, success, and failure. This all, naturally, leads to thoughts of winners and losers, and to what one can only describe as a somewhat male-oriented view of the world. Epistemically, I do not see these models as wrong – in fact they are most successful – but they do incorporate a certain non-epistemic value perspective.

McMullin might object here; but, rather than attempting an inadequate reconciliation, in too short a time, let me simply note why I feel uncomfortable with McMullin's claim that non-epistemic values will go as we develop and confirm the hypotheses of science. At root, I suspect that McMullin and I start from fundamentally different metaphysics, leading to most different conceptions of that 'reality' with which science grapples. McMullin I see as having a conception of reality relatively independent of humans – a kind of Kantian thing-in-itself, where trees exist in the forest when no one is about. Thus, for him, there has to be a goal – given the fact/value distinction – where, at the very least, all non-epistemic values are stripped away, and even the epistemic values are seen to be mere crutches to get at (non-value laden) reality.

For me, reality is meaningless, unless one thinks of it as involving the interpreting human. I deny idealism, in the sense of the world being but a function of ourselves, of our imaginations in some fashion. I am certainly no relativist, believing (no less than does McMullin) that the epistemic rules give us stability and sense of progress towards a shared understanding of an objective world. However, this understanding can never be one which eliminates the knowing subject. Thus, one expects the persistence of values, of all kinds, in our science.

As a Darwinian, I would try to link our knowledge of the world with our senses and powers of reason, as given to us by natural selection. But this is a story to be told elsewhere (Ruse 1986b). Here, I simply note that my reality – my objectivity – never gets away from the knowing subject. And, since ideological factors (like morality and religion) are so crucial to knowing subjects, it is for this reason, even in the very bones of science itself, that I side with the feminists in sensing the existence and importance of non-epistemic values.

CONCLUSION

Koertge's fictional Helen is too extreme in her view of science. It is not myth or something on a par with myth. But, non-epistemic values of the kind which excite feminists are important in science, and probably a great deal more pervasive and lasting than we conventional philosophers of science recognize and allow. We would do well not to ignore or deny feminist strictures, but to listen in a spirit of enquiry – a spirit which should be critical, but also friendly.

NOTES

1 I am not sure quite what Helen intends in the above-quoted passage. Does she think science and myth are different, but alike in being drenched in values? Or does she think that, essentially, science and myth collapse into each other? I suspect she would find my worry supremely unimportant.

2 Interestingly, McMullin himself would probably deny that Lyell was influenced by non-epistemic values, in the way I have discussed. At least, he would admit the influence, but deny that one can properly speak of 'values' here. If God is as the deists suppose, then this is a matter of fact not value. 'What I am arguing for is the potentially *epistemic* status that philosophical or theological world-view can have in science' (McMullin 1983: 19). In response, while I fully accept the fact/value distinction (Ruse and Wilson 1986), I suspect that there are borderline cases. Sickness is a fact, but as we describe it, it has non-epistemic value connotations. So also does God. Certainly, if God is all-perfect and all-loving, then we ought to worship Him.

3 Nagel (1961) rather dismisses the importance of one's choice of topic. He points out that choice is as crucial in the physical sciences as it is in the rest of science. To which I respond that I look to the significant importance of non-epistemic factors in physics, rather than to the general unimportance of non-epistemic factors in science.

4 We should not underestimate the importance of language in science. Central to the row after the *Origin* was published was Darwin's very term of 'natural selection' (Hull 1973a, Ruse 1979a).

Chapter Nine

IS RAPE WRONG ON ANDROMEDA?

An introduction to extraterrestrial evolution, science, and morality

One of the biggest money-makers in the history of the movies starred a little fellow with big appealing eyes, a fondness for candy, and an inability to hold his liquor. I refer, of course, to ET, the friendly being from outer space, who got left behind when his space ship lifted off prematurely. Audiences all over the world cried at ET's disasters, and cheered at his triumphs. He will surely some day soon get his own mark on Hollywood Boulevard.

ET is only one element in the general fascination we seem to have today with possible beings from outer space. Successful movie after successful movie plays on this entrancement, as also do some of the longest running television situation comedies.[1] Science fiction books and stories sell inordinately well and there is a constant demand for such old standbys as Superman comic books. On top of all this, government agencies spend our money on searching for possible life either here in our own solar system or further out in the universe. That such searches have so far been unsuccessful seems merely to be a stimulus for renewed efforts and renewed demands for funds.

This fascination with possible life elsewhere in the universe is not a new phenomenon. The ancient Greeks speculated on the subject, and down through the ages there has been a constant interest in the possibility of life elsewhere. There has also been speculation about its possible nature. Will it be more intelligent than we are? Will it worship the same God? And above all else, will it be friendly? (See Beck 1972, and Dick 1982, for historical references.)

In the past, philosophers contributed enthusiastically and significantly to the debate about the possibility of extraterrestrial life. Many of the greatest names in philosophy, for instance Aristotle and Kant, had things to say on the subject.[2] Some lesser writers dealt with the topic at great length. Perhaps the most detailed contribution to the extraterrestrial debate – one which denied the possibility of such life – came from the pen of the nineteenth-century British philosopher William Whewell, who argued vigorously that such life is not possible (Whewell 1853). In so doing, Whewell sparked off one of the liveliest controversies of the mid-nineteenth century, a controversy which raged until Darwin's even more shocking ideas diverted attention elsewhere (Brooke 1977, Ruse 1979a).

And yet, despite the history of philosophical interest in extraterrestrial life, and despite the general interest today in such life, modern philosophers stand curiously aloof from the whole debate. There is hardly anything written on the subject. In 1971, as President of the Eastern Division of the American Philosophical Association, Lewis White Beck (1972) used his presidential address to invite philosophers to rekindle their interest in the debate. However, his invitation fell on deaf ears. Somehow, philosophers are altogether too sober a group to discuss extraterrestrials. Dabbling with such entertaining beings as Superman or ET is too frivolous an enterprise for us. When we turn from serious discussions about the foundations of epistemology and ethics, it must be to even-more-serious discussions of such weighty topics as abortion and reverse discrimination. Science fiction is too childish a pursuit for modern philosophers to take seriously.

It is a pity that the philosophical community did not respond positively to Beck's invitation. *Pace* Dr Johnson, philosophy does not have to be serious and ponderous to be important. Even good philosophy can be fun. Moreover, looking at the extraterrestrial life question can pay philosophical dividends in its own right. At least, this is what I shall try to demonstrate in this essay, for I intend to share with you some thoughts about life elsewhere in the universe.

WHY CONSIDER EXTRATERRESTRIALS?

I must explain my own interests and the consequent direction my discussion will take. In a funny sort of way – particularly funny for someone who is about to write an essay on extraterrestrial life – I am not myself very interested in whether or not life really does exist out there in the universe. For William Whewell, the possible existence of extraterrestrial life, particularly extraterrestrial intelligent moral life, was rightly a matter of some personal concern. As a practising Christian, he felt that, were worlds other than ours populated with intelligent beings, then necessarily they would have to have some kind of relationship with God. And this would definitely dilute any special unique relationship that we humans could claim to have with God.

Indeed, the possibility of such beings elsewhere in the universe raised for Whewell all sorts of horrendous questions about the possible need for Jesus to come down from heaven and be crucified for their sakes, as well as for ours.[3] One has the dreadful thought that on Friday, every Friday, somewhere in the universe Jesus is being crucified for someone's sins.

Whewell's critics, who argued that there is indeed life elsewhere, had no fewer personal concerns than he. The most strenuous objector was the Scottish educator, scientist, and ardent natural theologian, Sir David Brewster. Author of a vitriolic, passionate work with the glorious title, *More Worlds than One: The Creed of the Philosopher and the Hope of the Christian* (1854), Brewster was desperate to show that no part of God's creation is without purpose. Thus Brewster felt that he had to prove that everywhere, throughout the universe, there is teeming life. This extended even to our sun, despite the fact that, heat problems aside, any human on the sun's surface would weigh somewhere in the region of two tons, because of gravity.

My worries are not the worries of Whewell and Brewster. My religious beliefs need no proof or disproof of the existence of beings elsewhere in the universe. Indeed, from a personal point of view, I am quite indifferent as to whether or not there is life elsewhere in the universe. I find this life of ours quite enough to handle! Furthermore, although I am sure that a philosopher could rightly enter the dispute about the probabilities of such life existing elsewhere in the universe, I shall not really do so in this article. Some well-qualified scientists today argue that not only is extraterrestrial life possible, it is indeed probable (Billingham 1981). Other equally well-qualified scientists argue that extraterrestrial life is most improbable (Hart and Zuckerman

1982). I suppose that if I were pushed, I would admit to a sneaking suspicion that, possibly, extraterrestrial life does exist. But, I must agree that often calculations of the possibility or probability of extraterrestrial life put one in mind of the engineer's way of solving problems: namely, think of a number, double it, and the answer you want is half the total.

All too often, in the face of ignorance, we make assumptions which sound reasonable, but which are totally without justification. Suppose you are told that there are some hundred million possible sites of extraterrestrial life in the universe. Suppose then the suggestion is thrown in that, on one in a thousand of these sites, such life has in fact evolved. That sounds reasonable enough. What is one in a thousand? But without evidence (usually lacking) it is quite without foundation, and why not one in ten, or one in ten thousand, or one in a million, or one in a hundred million? Unless one has either a theory to back up one's hypothesizing, or some empirical evidence to show that one's probabilities have some possibility of being reasonable, any supposed probabilities that one pulls out of thin air can be no more than that.[4] Thus, I conclude only that perhaps life exists elsewhere in the universe, perhaps it does not. Perhaps such life is intelligent, perhaps it is not. I simply do not know.

Why then, you might ask, should one professing so little interest, and even less knowledge, presume to write on extraterrestrial beings? The answer is that, by so writing, we can throw interesting light on ourselves. Moreover, egocentrically, I must confess to an extreme interest in human beings. Hence, my aim in this essay is to consider extraterrestrial beings – most particularly to consider their possible natures – hoping thereby to understand more fully the factors which make human beings what they are now. I am particularly interested in human claims to knowledge, both in the realm of epistemology (what Kant would call pure reason) and in the realm of ethics (what Kant would call practical reason).

So, supposing that there are beings elsewhere in the universe, the questions I ask are the following: How did they come to be? Was it through some form of evolution, and if so, what form? What are they like? How then do they compare in origins, similarities, and differences with us human beings? What does this tell us about ourselves? Specifically, what do we learn about

our intellectual and moral capacities, potentialities, and limitations?

EVOLUTION?

Let us assume that somewhere in the universe there is a planet
roughly like ours, with the approximate combination of
chemicals and conditions that obtained on our planet some
three and a half or so billion years ago, after it had cooled down
somewhat, but before life had yet appeared. I admit that I do
not know if this is at all a reasonable assumption. Since I am
carrying out a thought experiment, it is enough that the
assumption be logically possible.[5]

As a determinist and a mechanist, I certainly believe that,
given the identical combination of chemicals and conditions as
prevailed here on Earth, life would necessarily appear in exactly
the same way. But even thought experiments cannot ask this
much – at least they shouldn't. I'm just assuming that we have
more or less the same chemicals and conditions. And this being
so, we certainly cannot say definitively that life would appear on
our hypothetical planet. We have neither established theory nor
strong empirical evidence to let us say this much.

However, today, unlike just a few years ago, we are far from
completely ignorant about the coming into being of life.[6] We
have some very suggestive empirical evidence showing how the
components of life could be formed naturally from non-life
substances. I refer in particular to the work, in the 1950s, of
Miller, who showed that amino-acids can be formed from inorganic substances without direct human intervention. Furthermore, a number of very stimulating theories have recently been
developed about how life could develop naturally from these
components (Dickerson 1978, Schopf 1978). For instance the
Nobel prize-winner Manfred Eigen has devised a theory showing
how ribonucleic acid (RNA) could be formed without human or
other intelligent direction (Eigen et al. 1981).

Hence it is no longer true to say that we are totally without any
knowledge whatsoever about life appearing naturally. Nor is it
reasonable today to conclude that, given a planet much like ours,
life could never have appeared. But again, I will not become too
bogged down with facts! If you think that, given a planet like

ours, life would not appear, then imagine a planet like ours with life. (With my cavalier disregard for facts, I am even less concerned with whether life could have occurred under very different conditions and combinations of chemicals: for instance, whether there could be some sort of ammonia-based life. If indeed such alternatives are real possibilities the points I want to make will still hold valid; if they are not real possibilities, we have not been side-tracked by them.)

Let us suppose then that life does develop in some primitive sort of form. I am deliberately avoiding heady philosophical discussions about what we mean by 'Life'. I am simply assuming, backed by suggestive experimental and theoretical work mentioned just above, that we have primitive organisms which are relatively stable. Moreover, and this is crucial, they *reproduce.* They do not just sit around in splendid isolation, contemplating their DNA. They multiply.

At this point, big, obvious questions intrude. Will there be evolution? And if so, what will fuel this evolution? Here on Earth, we do have evolution, and *the* major mechanism of change is 'natural selection'. More organisms are born than can possibly survive and reproduce. In the consequent 'struggle', some prove more successful or 'fitter', by virtue of their special characteristics. This leads to a kind of natural winnowing or 'selection'. Given enough time, evolution occurs. But would the terrestrial mechanism of natural selection be as important on our hypothetical planet, as it undoubtedly has been down here on this real planet?

There will be evolution, and natural selection will be crucially important. Why can one say this? The answer lies less in evolution itself, and more in the natural selection which will in turn lead to evolution. Natural selection is not a tautology, as some critics claim.[7] However, it does rest on basic properties of organic matter, and, under the properties already proposed for our primitive life, it is hard to suppose that it would not apply up there to our hypothetical planet.

Consider: we are assuming that these are organisms which do reproduce, and I take it we are assuming that they do not simply reproduce one of their kind and then die out. Rather they are organisms which multiply. Obviously, there is going to be an explosion in numbers, as there is a Malthusian geometric increase. Eventually, fresh living space and additional nutrients required for survival will be exhausted. What then? On our hypothetical planet, there will be a struggle, as there is down here on our real planet. If all the organisms are absolutely identical, then success will be a matter of chance. There will be no selection. But, if there are differences, and everything we have learnt about life here on earth suggests there will be differences, then the slightest edge will lead to success in the struggle. There will be natural selection on our hypothetical planet.

Moreover, this natural selection will lead to evolution. That is to say, there will be an ongoing change in the organisms on the planet. This is not to say that all of the organisms will be changing all of the time. There could be long periods with little or no change, as occurs on earth (Stebbins and Ayala 1981). But, given enough time, one expects change to occur. This is an expectation backed both by theory and by terrestrial empirical evidence (Ruse 1982c).

BUT WHAT DOES IT ALL LEAD TO?

What route would evolution take on a hypothetical planet, and what effects would it lead to? In many respects, it is simpler to say what sorts of things we should not necessarily expect to find. Most particularly, there is absolutely no reason to think that there would be inevitable progression from primitive forms to intelligent beings. Life first appeared here on earth about three and a half billion years ago. But until just over a half billion years ago, such life was very primitive. It is only in the last 600 million years or so, that we have had the development of sophisticated organisms and plants (Valentine 1978).

Hence, if anything, the empirical probability seems to be that life on our hypothetical planet will remain at a very crude level. Success in the struggle for existence is not necessarily a question of being brilliant. Rather, it is a question of being a better reproducer than your neighbour. If your neighbour is burdened with brains, whilst you are little more than a fancy sperm or egg sac, then it could well be that you will reproduce far more efficiently than he, she, or it.

Moreover, although a number of hypotheses have been put forward, no one has yet given absolutely convincing reasons why, at the beginning of the Cambrian or thereabouts (i.e. 600 million years ago), there was an explosion in life forms, including sophisticated life forms (Raup and Stanley 1978, Futuyma 1979). Hence, we have neither theory nor empirical evidence to suggest that life will necessarily evolve into something more than primitive single-celled organisms – however long a planet may exist. There is certainly no guarantee of birds and fish and reptiles and mammals and plants and everything else that makes life recognizable for us. There is absolutely no guarantee of races of ETs or Supermen. If anything, we should expect the persistence of low-grade life.

What then can we say about the effects of natural selection on our hypothetical planet? The most important thing about natural selection is that it does not just lead to any kind of change. It brings about adaptation. That is to say, natural selection makes for characteristics which aid their possessors in life's struggles. Moreover, these are characteristics which help each organism individually. Such features are known as 'adaptations'. Hands, eyes, ears and noses are all organs which help one to survive and reproduce. These are all adaptations (Williams 1966, Lewontin 1978, Mayr 1982).

Hence, on our hypothetical planet, we expect that natural selection will bring about adaptations. Moreover, if most people have got one particular characteristic, helping survival and reproduction in one particular way, it often pays to have different characteristics, helping survival and reproduction in other ways. Natural selection usually leads to diversity. If everyone is trying to eat one foodstuff, then, if there is another available foodstuff, a liking for this second foodstuff is clearly to one's advantage. I expect, therefore, that on our hypothetical planet, we would find different forms of organism (Dobzhansky 1970).[8]

Whether we should find sexuality on our planet is an interesting question, but I am not sure that we can really answer it that definitively. There are almost as many theories on the evolution and maintenance of sexuality, as there are writers on the topic (Ghiselin 1974a, Williams 1975, Maynard Smith 1978b, Trivers 1983). There does seem, however, to be strong evidence to suggest that sex has evolved more than once here on earth, so perhaps it might evolve elsewhere also. If indeed there is sexuality, then probably there should be species: groups of reproductively isolated interbreeding organisms. There is still much controversy about how speciation actually occurs, but it does seem to come about as a consequence of sexuality. As soon as organisms start breeding with one another, then things happen which stop that breeding from being ubiquitous (Dobzhansky 1951, Mayr 1957, 1963, Grant 1971, 1981, Sokal 1973, Van Valen 1976, Endler 1977, White 1978, Levin 1979).

Let us move on now towards questions about human evolution. I have said that there is absolutely no guarantee on our hypothetical planet of an upward progression to intelligent life forms. On the other hand, evolution is constantly trying to exploit new unused domains. As one area, for instance the sea, becomes used up and occupied, then evolution pushes organisms into unused areas, such as land, and as that is used up, then evolution turns towards other areas, such as the air (Valentine 1978, Feduccia 1980). And similarly, in our own case, we have seen an evolution from a more biologically oriented sphere to one which involves more of a cultural dimension, and which is less connected absolutely with brute nature. There was nothing inevitable about this development, but it did happen nevertheless. Furthermore, notwithstanding all of the reservation, intelligence is not something terribly counter-intuitive, given natural selection. One simply has the constant need to better oneself, and bettering oneself involved becoming a little more sophisticated. Intelligence is the epitome of sophistication (Johanson and Edey 1981, Lovejoy 1981).

What I suggest, therefore, is that through a process of natural selection, we might possibly have the evolution of more sophisticated organisms on our hypothetical planet, and these could even reach the realm of intelligence. Since it has happened once, namely here on Earth, I do not think it is empirically impossible that it happened elsewhere. But I do emphasize that such evolution of intelligence is not a necessary consequence of life appearing: not at all. (One important question, which I will leave undiscussed for the moment, concerns possible connections between sex and intelligence. I will look at it later.)

Let us suppose that natural selection does lead to some sort of intelligent race or species of organism on our planet. (At least once here on Earth, we had coincidentally two species of semi-intelligent organism: *Australopithecus africanus*, and *Homo habilis.* Members of neither species were as intelligent as members of *Homo sapiens* are.) The question which arises now is just what sorts of knowledge systems would these extraterrestrial intelligent beings have? In the next section, I turn to the question of epistemology. That is to say, what sort of knowledge would they have of the world around them? For instance, would they know of Newtonian gravitational theory, and of like discoveries of the human mind?

EXTRATERRESTRIAL EMPIRICAL KNOWLEDGE?[9]

There is an obvious fallacy, even if an understandable one, which occurs in popular science-fiction movies. This is the making of the extraterrestrial beings as recognizably human, particularly with respect to their senses. Admittedly, Superman had supposed powers which are beyond us; but generally the extraterrestrials are given those very senses which we have, often with the same intensities. Thus, for instance, ET had two eyes, just like humans, as well as a human sense of taste (otherwise, why was he so fond of Reese's Pieces?), together with a human sense of touch and smell and so forth.

But we can certainly say from our experience here on Earth that there is no reason at all why an intelligent denizen of our hypothetical planet should have precisely those senses that we have, in precisely the way that we have. For instance, certain snakes can strike their prey in the dark, by virtue of having extremely sensitive heat sensors on their heads. I see no reason at all why an extraterrestrial being should be blessed with a human sense of sight, rather than with sophisticated heat sensors. Again, there are more ways of seeing than simply the human ways. Spiders and some other animals have evolved a sense of sight quite independently from the mammals, so it could be that our imaginary beings would see parts of the spectrum different from those which we see.

It is well known too that many animals – fish, insects and others – communicate by chemical means, through so-called 'pheromones'. To give an idea of how efficient this method of communication might be, I quote from a discussion by E.O. Wilson, an expert in the nature and use of pheromones in the organic world:

The amount of potential information that might be
encoded [by the use of pheromones] is surprisingly high.
... Under extremely favourable conditions, a perfectly
designed system could transmit on the order of 10,000 bits
of information, an astonishingly high figure considering
that only one substance is involved. Under more realistic
circumstances, say for example in a steady
400-centimetres-per-second wind over a distance of 10
metres, the maximum potential rate of information transfer
is still quite high – over 100 bits a second, or enough to
transfer the equivalent of 20 words of English text per
second at 5.5 bits per word. For every pheromone released
independently, the same amount of capacity could be
added to the channel capacity. (Wilson 1975a, p. 233)

Clearly, we humans hardly begin to exploit all the ways in which
information can be handled. To think that extraterrestrials will
use all, and only our, five senses is unjustifiably anthropomorphic
in the extreme. Intelligent denizens of our hypothetical planet
could be blind snakes, rather than human-sensing fellows like
ET.

I suppose, however, that if extraterrestrials are to do anything
with their intelligence, they will have to have some ability to
manipulate their environment. Perhaps indeed, intelligence and
an ability to manipulate the environment go hand in hand. (If
you will forgive the metaphor.) Certainly, the history of human
evolution seems to imply that physical and mental evolution are
connected, although now it is believed that we pushed towards
our present physical state somewhat before we pushed towards
our present mental state (Johanson and White 1979, Johanson
and Edey 1981). But I do not see in principle that natural
selection lays down absolute rules for the way in which the
evolution of intelligence must take place. It is simply that in
order for the evolution of intelligence to occur, the individual
must in some way be able to put it into practice. We therefore
look for feedback between the ability and the means of
utilization. A human being is not going to develop straight into
a philosopher, able only to think and not to act.

Will our extraterrestrials be conscious?[10] I am not quite sure
how you would answer this question in any definitive way. We

know, each and every one of us, that we are conscious. We are as sure of the consciousness of our fellow human beings as makes no real difference. We are equally sure that stones are not conscious and fairly certain that lower organisms like oysters have little by way of consciousness. There are those who have denied that any animals, other than humans, have consciousness. René Descartes was one. I suspect, however, that most people would agree that consciousness is at least in some sense linked with brain power, and that it is shown through physical actions, behaviours, responses, and so forth. This being so, most people would allow that the very highest organisms, (other than humans) have some rudimentary form of consciousness.

Certainly, it seems a little odd (selfish?) to deny that chimpanzees and gorillas have consciousness of any kind. They engage in behaviour which goes along with what we associate with consciousness in our fellow humans, and indeed in ourselves: gorillas and chimpanzees are able to manipulate tools, make discoveries, behave in social ways, and so forth. They do not have verbal speech; but it is arguable that actually being able to talk is a necessary condition for any form of consciousness whatsoever, especially since there is increasing evidence that the higher organisms can communicate in non-verbal ways (de Waal 1982).

Let us grant that animals other than ourselves are conscious. At least, let us grant the weaker premise that an organism as sophisticated as we are in brain power could be conscious. Hence there is no good reason to deny the possibility of consciousness in extraterrestrial beings. However, oddly, I am not sure whether the possible consciousness of our extraterrestrials is all that important, or all that significant. At least not for us! I see no reason why extraterrestrials should not behave and react in just such a way that we would associate with reasoning, whether or not the kind of 'ghost in the machine' so beloved of dualists actually exists. Put matters this way. In the *Star Wars* series, C3PO and R2D2 are machines. Luke Skywalker and Princess Leia are human-like. Chewbacca and Yoda are intelligent non-humans. Does one want to say that any or none of these is conscious? Does it matter?

We come now to a second, and rather more interesting, question. What sort of thinking beings would our extraterrestrials be? What sort of consciousness would they have, if they had consciousness? Clearly, in many important respects, our extraterrestrials are going to think in ways very different from ours. At least, we cannot expect that they will not. For instance, if their senses are essentially chemical-using, then they are going to have a whole range of sensations totally unexperienced by us. Conversely, the sensations experienced by us are going to be unexperienced by them (Grossman 1974).

At the very least, what our extraterrestrials take for granted will be theoretical concepts for us, and vice versa. Their metaphors, similies and all of those sorts of concepts would be quite different. One should not underestimate this. There seems to be something of a tendency to think that analogies and metaphors are merely useful aids, but theoretically dispensible; that they are not things which enter into real thought, in particular into science. However, in common with a number of other philosophers of science, I would dispute this. Metaphor plays a very important role in science, and to eliminate it would be to alter science very drastically (Black 1962, Hesse 1966, Ruse 1973a, 1973b, 1977a, 1980b, 1980c, 1981b, 1981c).

Indeed, I am not at all sure that one could eliminate metaphor entirely from science. Therefore, to say that our extraterrestrials could grasp what we would think of as Newtonian mechanics is dubious. For instance, for us an important part of Newtonian mechanics is a notion of 'force'. This is clearly something we get through the idea of actually physically stretching and pressing, and so forth. However, if one did not communicate through touch at all, or at least only in a very minor or unimportant way, but rather through some sort of chemical sensing, then I doubt that the notion of force could play the same role in extraterrestrial mechanics that it plays for us. Conversely, some other metaphorical notion would be at the centre of their theory. That is, assuming they have an equivalent theory.

Again, in company with other philosophers of science, I take the notion of a 'problem' to be very important (Laudan 1977, Ruse 1980b). Scientists do not just go out and discover the world – any old part of the world. They get stimulated by problems. Newtonian mechanics is obviously a result of our awareness of the heavens and of gravity (to name but two factors). What if we were blind fish? Would we feel the need for Newtonianism? Certainly not in the same way. Astronomy would be less important, hydrodynamics more so.

BUT IS THERE NOT SOME UNIVERSAL KNOWLEDGE?

However, you might think that, despite all differences between extraterrestrial science and human science, there would nevertheless have to be a basic underlying similarity between any kind of human thinking and any kind of extraterrestrial thinking. After all, $2 + 2 = 4$ on (a planet in) Andromeda, no less than it does here on Earth. Squares are squares and not circles on Andromeda, just as they are here. And, whether or not we can or do think of 'force = mass x acceleration' on Andromeda in human terms, it holds up there equally as much as it does here. And the same is true of Boyle's law, Snell's law, and all the rest. Furthermore, if, as I have just been suggesting, extraterrestrials evolved through natural selection, then natural selection is no less a reality on Andromeda than it is down here on Earth.

Hence, whether or not the periphery of science is changed, there is something underlying which is the same. There is, therefore, a universality to what can and will be known, despite the differences. The situation is analogous to that seen by Chomsky (1957) in linguistics, when he argues that there is a universality to language, a so-called 'deep structure', despite the many surface differences there are between (say) English and Japanese.

Two comments can be made about this point. First, even given the amount that the critic would allow, I think it would be terribly easy to underestimate the grave differences which could still exist between us and extraterrestrials. We could have to battle all the way for even the most primitive communication. We know full well how difficult it can be to understand the minds of people in cultures different from our own. How many Westerners, for instance, would truly be able to say that they could understand oriental inscrutability? But, if there are so many difficulties encountered in communicating with, and fully understanding, our fellow humans, think how much more difficult it could be to try to communicate with extraterrestrials.[11]

Certainly any kind of profound understanding of extraterrestrial culture could be forever out of our grasp. Conversely, any understanding of us could be out of the extraterrestrials' grasp. And do not think that breakdown in communication would occur just in the arts, although how one would explain Wordsworth's poem about daffodils to a blind watersnake I do not know. As I have argued above, we could have a great deal of difficulty communicating very many scientific ideas, given the extent to which they are permeated with our human values and nature.

The second comment is that, despite all difficulties of communication, I think that there would remain some common ground in human and extraterrestrial knowledge. Essentially, what I am arguing is that there is a common world out there, one to which we and our extraterrestrials will have adapted. Without such an assumption, everything I have said so far falls away. Hence I am committed to a realist position. (The sense of 'realism' will be spelt out shortly.)

The world is not just a figment of our imaginations. Rather, we are organisms who have adapted, through natural selection, to deal with external realities. Moreover, I have no reason to doubt that really 2 + 2 does equal 4. Why? Because I have no reason to, that's why! It is simpler to assume that a being which recognizes that 2 + 2 does equal 4 is better adapted than one which does not, if indeed 2 + 2 equals 4 and not 5.

Again, I have no reason to believe that circles are not really circles and are really squares. Why? Because, it seems more sensible to assume that an animal which has developed adaptations enabling it to deal with circles which really are circles rather than squares is in some sense better off than an animal which has not. Furthermore, I see no reason to think that on Andromeda, or anywhere else, 2 + 2 will not equal 4, or that circles will really be squares, or that the fire will not cause burning, or that more generally the same cause will not lead to the same effect. And I see no reason to doubt that Andromedans will be better adapted if they (like us) can distinguish circles from squares, and can learn that fire 'causes' burning. (But note what I have to say later about causation.)

In short, I am assuming that there is going to be an underlying similarity between the world humans live in and any world elsewhere. Additionally, extraterrestrials, just like humans, will have learned to cope with their world and to recognize the underlying patterns which are common to the whole universe. I would even go so far as to say that if there are good adaptive reasons for believing in the existence of a god, namely that it promotes a certain group stability and cohesion, then perhaps extraterrestrials will believe in a god (Wilson 1978). But remember the numerous varieties of beliefs here on Earth, with the Hindu notion of God bearing but the faintest family resemblance to the Christian one. Even if we met extra-terrestrials, I am not at all sure that we would be able to under-stand what their notion of God was, or even that they had one.

My key position, therefore, is that there will be some underlying basic premises held in common by humans and extraterrestrials. I have in mind such things as the basic premises of logic, of mathematics, of causality, etc. In other words, I am thinking of the kinds of concepts that Kant identified as analytic, and perhaps also the synthetic a priori. Thus far, extraterrestrial epistemology will be similar to human epistemology. (Again, note the qualification I make later, at the end of the next section, about the 'reality' of such things as causes.)

ULTIMATE REALITY?

There is a major question which must be raised and looked at squarely. I am suggesting that organisms are a product of natural selection, and that they have senses because these are adaptive. What this means is that having senses enables organisms to survive and reproduce better than if they did not have them. Clearly, this is no guarantee of ultimate truth in any sense. It is rather a pragmatic position (Quine 1969). We have certain senses because these work. But, since humans are themselves organisms, cannot the argument be turned back onto everything that I have just been saying? What guarantee have I that any claims I make about natural selection itself are really true? What guarantee is there that my claims really tell us of that which is out there, be-yond human fallibility and adaptive need? Is natural selection absolutely true of ultimate reality, even though we ourselves learn of it through organs themselves fashioned by selection? Indeed, can we properly talk about ultimate reality, as opposed to the common-sense reality we experience all the time – an everyday reality to cope with which selection gives us adaptations? To use the terminology of the philosopher Hilary Putnam, can we talk of 'metaphysical reality' as opposed to 'internal reality' (Putnam 1981)?[12]

Put matters this way. My human senses, my human powers, are themselves products of natural selection. Therefore, everything that I am saying could be totally mistaken (Trivers 1976). My claims could be simply things I believe, because I am a more efficient reproducer if I believe them, rather than otherwise. In short, apparently, ultimately I have no right to say even that 2 + 2 really equals 4 or that there really are chairs and tables where I think there are, or that the fire always causes pain, or anything else. I think these things hold true because these beliefs enable me to survive and reproduce. But what ultimate reality is like, I cannot say. Nor – and this is the punch-line – can I say that ultimate reality for extraterrestrials will be the same as it is for us.

This is a worrisome objection, but we can go some way towards answering it. Moreover, the distance that we can go provides all the answer that we really need. First, note that we are not entirely helpless or uninformed about the world out there. Certainly, we are not uninformed about the effects of different senses, or of senses alien to us. At the personal level, we ourselves have several senses, and the interesting thing about these senses is that, although they give different information, they do nevertheless cohere on the reality of the world.

I think, in particular, of the sense of sight and the sense of touch. It is well known that the proper objects of sight and the proper objects of touch (as Berkeley used to call them) are not the same. A blind man who knows a cube from a sphere by touch, cannot immediately distinguish a cube from a sphere (or a square from a circle) when once he is given sight.[13] Nevertheless, the blind man can distinguish the cube and the sphere just as the man with sight can. He knows that he has two different objects, and that the one is what he would call a 'cube' and the other a 'sphere'. Moreover, these are just as stable and usable as are the 'cubes' and 'spheres' of the man of sight. Hence, although different senses give us different information, they do cohere on such things as differences between cubes and spheres, as well as on 2 + 2 = 4. Therefore the fact that we have different senses does not mean that we cannot or do not agree on the same basic underlying principles.

Moving now from ourselves to other animals in this world of ours, note that chimpanzees seem to avoid objects in their path, just as much as humans do. (Rather better, in fact!) Similarly, inasmuch as chimpanzees can manipulate they do it in essentially the same sort of ways that humans do and for the same sort of reasons. Again, moving now beyond straight sensation to the ways in which we process it, we find that chimpanzees learn to associate cause and effect, in the same ways that humans do. For instance, a fearful sound sets up an expectation in a chimpanzee just as it sets up an expectation in the human (de Waal 1982).

What I argue, therefore, is that we do not get a total division between different organisms, or between organisms of different species, with respect to the fruits of sensation. Nor do we get total breakdown with respect to the basic principles that I suggest might underlie all epistemology, including human epistemology. This is not to say that chimpanzees or lower organisms are as aware of these principles as humans are. In fact, I suspect that we are the only organisms here on this Earth consciously aware of the kinds of underlying principles I am talking about. But what I am saying is that they do seem to be shared by organisms of different species.

Even when we move to organisms which use entirely different senses, like those organisms relying on pheromones, we still find that there seems to be the same basic consistency between them and us, and other organisms using senses like ours. For instance, snakes using heat sensors strike at objects which we can see. They do not strike at invisible prey, and then slither through that which we sense as solid. And the same points hold for animals using pheromones. We see the objects or we feel the objects; they sense them chemically or some such way (Wilson 1975a).

In other words, there seems to be a basic underlying consistency with respect to reality, and to the primary properties of reality, which transcends different senses. There seems, therefore, no reason to deny that the same should be true of our extraterrestrial beings. Although their ways of sensing could be quite different from ours, the evidence from ourselves and from our fellow Earth-organisms is that reality is the same for all, and that our various senses are all attuned to react to its fundamental properties. Hence, at this level, one can say confidently that extraterrestrial beings will share with us certain fundamental, underlying principles. Whether they will have the precise awareness of them that we have is another matter. But the principles will be shared nevertheless.

Having said this much, however, there is still the underlying worry that, in some ultimate way, we may just be entirely mistaken. What is shared by humans and extraterrestrials is still the reality revealed by the organs of selection (Putnam's internal reality). If, as I argue, our senses and powers are adaptations to deal with the world, the critic might yet object that there is no absolute guarantee that we are not systematically deceived about some aspects of ultimate reality (Putnam's metaphysical reality). Descartes supposed an evil demon who misled us. Perhaps natural selection misleads us 'for our own good' (Trivers 1976).

At this point, I simply have to throw up my hands and admit that one can set up the situation in such a way that, ultimately, I have no answer. But, note that this supposition says nothing to refute what I have just argued about the similarity between ourselves and extraterrestrials. I see no reason to believe that Descartes' evil demon would not deceive them as well as us, or that they would not be subject to the same kind of natural-selection-caused distortions about ultimate reality that we are.

However, I do not want to end on a negative note. The reader should not think that, because I admit we are limited, we can say nothing about ultimate reality, and that such ultimate reality could be of a nature that, if perhaps we only had extra senses, we could find out more, and that we could perhaps find out that $2 + 2 = 5$, and why it is biologically advantageous for humans to believe $2 + 2 = 4$. We should not think that perhaps extraterrestrials do have the needed extra senses, and thus can find out more than we can, even if they too in the end are necessarily limited by their abilities.

To counter this worry, let me say the following. First, there is something awfully fishy about this line of argument, if only because we are presupposing the truth of natural selection to prove that it might not be true. If in fact the real world is in no wise as we picture it, then it hardly makes sense to say that natural selection is systematically deceiving us for our own good. All we can do is remain silent before the imponderable – or, more precisely, before the inconceivable. To say that we are totally deceived by reality, just does not make sense. This is not to say any 'deception' is ruled out. I suppose, for example, we might think a certain colour identical with another – although I am not quite sure why this should be a product of selection and not simple insensitivity. Perhaps, more plausibly, we might say that we could all be deceived about the existence of God – although please do not take this comment as a plea for atheism. Apart from anything else, if it were argued that the use of the term 'deception' stretches metaphor too far in this case, I would be quite sympathetic.

Second, backing up the first point, because our powers are limited, or rather, because our powers are what we have and just a product of natural selection, it does not follow that it is reasonable to suppose that reality is other than what we think it is. To say that our powers are limited and that we can say no more is just to say that it is inconceivable that ultimate reality is other than what we think it is. And to say that things are inconceivable is not in any sense to say that they are reasonable or probable or another option. They are literally inconceivable, and one can say no more than that. Hence, I find myself untroubled by the thought that ultimate reality might be other than it is, or that extraterrestrials might have deep insights into this reality – insights hidden from us. They might perhaps sense colours that we cannot sense. I doubt that they think $2 + 2 = 5$, either because it is true or because natural selection finds it worth while systematically to deceive them. And, even if it did, I am not sure how we would ever find out.

Concluding this part of the discussion, therefore, I argue that there could well be radical differences in the epistemology of humans and extraterrestrials. But inasmuch as we can both think about the world at all, there will be certain similar underlying thought patterns. And we share our powers of comprehension, because we are both products of the same process of evolution. Were a Humean to object to my analysis on the ground that, although objects may exist 'out there', such things as causal connections do not refer to real powers, but are mind-supplied to deal with reality, I would feel sympathy. However, even if we only have constant conjunction in reality (and I am inclined to think that is all we do have), my position is simply that thinking in terms of cause and so forth is an adaptation for dealing with reality (imagine not associating fire with burning) and it is inconceivable that extraterrestrials would not share such adaptations.[14]

You may think that, in this sense, selection is 'deceiving' us. But apart from the fact that David Hume was not deceived, I have already hinted that emotive words like 'deception' are probably not really appropriate here. We are not consciously being led into falsity. (I would like to think I am arguing for a Kantian solution to Hume's problem, but one which is infused by Darwinian insights.)[15]

EXTRATERRESTRIAL MORALITY

I come now to the most interesting questions of all. What about right and wrong, likes and dislikes? Will inhabitants of our hypothetical planet have any kind of moral code? If so, will their code be in any way anything like our own? And, most particularly, how would such extraterrestrials feel and behave towards us humans if we were ever to meet them? Perhaps I should also ask what kinds of moral obligations we humans would have in return towards extraterrestrials.

If you are Christian or a Platonist, for example, then you believe in an absolute moral law. This being so, then whether extraterrestrials exist or not makes no difference to the ultimate moral rule for us all, whether we live here on this planet or somewhere else. However, if you are like me, you are nothing like as convinced of such ultimate rules. Matters therefore become rather more complex and more problematical. And, of course, even if you believe in an absolute moral code, you still have questions about the reactions that beings very unlike ourselves would have towards this code, and the obligations which would extend to them.

What follows if you take the kind of approach that I have been taking in this essay? What happens if you think that evolution through selection will be the causal key to intelligent life elsewhere in the universe, as well as here on Earth? Until recently, I think most people would have felt that there was little, if anything, that one could say about morality. There was a strong feeling, even by evolutionists themselves, that evolutionary thought, particularly evolutionary thought centred upon natural selection, had little or nothing to say about the evolution of morality. It was felt that evolution promotes little more than an unbridled selfishness where all organisms fight for their own interests, entirely to the detriment of any other organisms. In other words, there is an unrestricted 'nature red in tooth and claw'. This is social Darwinism, where the weakest go to the wall and the strongest triumph (Dobzhansky 1962, Flew 1967, Huxley 1894).

This is obviously all directly opposed to any kind of morally governed society, where kindness and friendship and altruism exist. Most people, therefore, including evolutionists, felt that somehow morality was a function of non-biological culture or some such thing. Morality and evolution are opposed, not complementary or in any way in harmony.

Recently, however, there has been a revolution in evolutionary thinking about the beginnings of morality. More accurately, we should say that there has been a return to the thinking of Charles Darwin himself, for in the *Descent of Man* Darwin was quite adamant that conventional morality can be given a selectionist backing (Alexander 1971, 1979, Trivers 1971, Wilson 1978, Ruse 1980a, Murphy 1982, although see also Singer 1981).

There are a number of causal models which are suggested in explanation of the way or ways in which our sense of morality could have evolved. One of the most developed is that which applies to relationships within the family. It is argued that there are good evolutionary reasons why people should behave in an altruistic, moral way towards relatives. In particular, we all share the same basic units of heredity, the genes. Thus inasmuch as one helps relatives to reproduce, one is thereby helping oneself to reproduce, by proxy as it were. This mechanism, which has been labelled 'kin selection', is thought to be responsible for a great deal of the altruistic behaviour which we show towards the people around us, for instance, our children. Obviously, help to relatives does not have to be confined to the actual mechanics of reproduction. Financial and other indirect support is covered by kin selection (Hamilton 1964a, 1964b, Trivers 1972, Dawkins 1976, Maynard Smith 1978, 1982, Oster and Wilson 1978, Wilson 1978).

What about dealings with non-relatives? Here also evolutionists (of the Darwinian ilk) argue that natural selection is important. In particular, it is argued that much morality evolved as a product of a kind of enlightened self-interest, or, as evolutionists call it, 'reciprocal altruism' (Trivers 1971, Alexander 1979). It is suggested that, inasmuch as one behaves in a friendly altruistic way towards other human beings, so also will they be prepared to behave in such a way to you. Moreover, those humans who are helped by others are in a better position to survive and reproduce than those who are not. This is so, although sometimes the recipients of help put themselves out for other people.

You might object, with Immanuel Kant, that true morality occurs only when one is a totally disinterested participant. That is to say, morality begins only when one does something without any hope of favour or reward. However, evolutionists anticipate this objection (Alexander 1974, 1975, Ruse 1979b). They argue that the evolved sense of morality in humans indeed does not necessarily involve conscious manipulation or calculation of possible return. In fact, evolutionists argue that one will probably function most efficiently when one has no hope of return at all. In short, biological functioning is enhanced when one is acting solely because one thinks that one ought to act in such a way. In other words, one works most efficiently when one is acting morally.

On the basis of the above and related models, evolutionists today argue that the human sense of morality is something which is a product of evolution, no less than the hand or the eye. Our sense of morality is an adaptation. We survive and reproduce more efficiently with it than we do without it. In the past, those people who lacked a sense of morality, simply tended to be ostracized and at a disadvantage. So they failed to survive and reproduce as efficiently as those with a sense of morality. This is not in any way to say that that which has evolved is morally good. In other words, this is not some form of crude social Darwinism, either endorsing the process of natural selection as a moral good, or the products of natural selection as moral goods. It is not to say that nature red in tooth and claw is an unqualified good. It is rather to say that the sense of morality which we have is a product of evolution.

Now, how does this all bear on the question of extraterrestrials? It should be remembered that we have concluded that extraterrestrials will, like ourselves, be a product of a process of natural selection. My surmise is that the answers will come at two levels, just as our answers to extraterrestrial epistemology came at two levels. Moreover, I think that the two levels answering to morality will bear significant similarities to the two levels answering to epistemology. There, it will be remembered, it was suggested that sensations, models, metaphors, analogies will probably be quite different for extraterrestrials and humans. However, at another perhaps deeper level, there will be significant similarities, particularly with respect to mathematics, logic, principles of causality. I would suggest that in the realm of morality, at a surface level, we will find very significant differences. But perhaps, at a deeper more basic level, there will be similarities between extraterrestrial morality and human morality.

SUBJECTIVE ELEMENTS IN MORALITY

At the top surface level, in human morality you find all sorts of subjective and variable elements. For instance, we find many differences between different societies. Take eighteenth-century India: people in that time and place practised and strongly endorsed suttee, in which widows would voluntarily perish in the funeral pyre of their husbands. Obviously, such a practice is totally alien to Western customs and morality. In fact, we think that widow sacrifice is totally immoral.

Again, we find differences between practices in Africa today and what we think moral in North America. Recently, there has been some controversy about the practice of female circumcision. Inhabitants of many parts of Africa think that this is a good thing to do, and that one ought to do it if one is to be a morally respectable member of society. We in North America and Europe, think to the contrary that it is a quite revolting and totally immoral practice, and see it as a way of unfairly keeping women in a subservient role.

Clearly, there is nothing particularly objective about this kind of morality, nor is it something that one would expect to find the inevitable product of natural selection. Even if one thinks there is an indirect connection (and I could think up a scenario to do with religion or sex roles), no one would pretend it is something we would expect to find practised among our extraterrestrials. (These practices presuppose sexuality, the possiblity of which in extraterrestrials will be raised later.)

I am sure some readers will be upset at the moral relativism apparently endorsed in the last two paragraphs (see Taylor 1978). It will be argued that suttee and female circumcision are objectively wrong, even though they have been widely endorsed and practised. They are not – and never could be – subjectively right. I have a sympathy with this objection. My point is that people in different societies have taken different things to be right and wrong. Often, although not necessarily, these relative rights and wrongs have been bound up with religious practices: I think, for instance, of food taboos and initiation rites. But if we argue – as I probably would – that, overall, religion and surface morality can have biological value, as with the powers of sensation, there is clearly more than one way to survive and reproduce (see van den Berghe 1979). Hence, at this level I expect our extraterrestrials to differ from us.

THE DEEPER LEVEL OF MORALITY

But now let us turn to the deeper level of morality. This is the level which corresponds, as it were, to the epistemological level of the rules of mathematics and causality: the level which informs and gives structure and meaning to our higher level judgements. Here, I see a great deal more universality and common acceptance of moral norms across *Homo sapiens*.[16] Moreover, as intimated above, at this level there is a much more direct connection between moral capacity and adaptive abilities as fashioned by selection. Hence I have no reason to doubt that some sort of morality would evolve between our extraterrestrials. After all, they have to get on with each other, just as we have to get on with each other. (This assumes that they are social: more on this later.)

I have suggested that the extraterrestrials will recognize 2 + 2 = 4. Will they recognize that which we would think of as the ultimate moral norms? My suspicion is that, in some way, they will. Two of the greatest and most widely accepted enunciations of the supreme principle of morality are the Greatest Happiness Principle and the Categorical Imperative. The former specifies that one's actions ought to be such as will maximize happiness (Mill 1910). The latter, due to Kant, has a number of formulations, but perhaps the easiest to grasp is that which entreats one to regard one's fellow humans as ends, and not simply as means to one's own gratification (Kant 1959). Either or both of these could find their equivalent on our hypothetical planet elsewhere in the universe.

Isn't this just wild speculation? Yes it is – but not that wild! Take the Greatest Happiness Principle. I do not know in what extraterrestrial happiness will consist. Perhaps it will be wallowing in mud baths; but, presumably, extraterrestrials will want to achieve their kind of happiness as best they can. There are good evolutionary reasons for liking sugar, sex, and sleep (Barash 1977, Symons 1979). There are even good reasons for the enjoyment of mental effort. Consequently, I see no reason why it should not be the case that there will be some kind of mutual feeling between the extraterrestrials, to the effect that, inasmuch as one maximizes the happiness of others, this is a good thing. This feeling, although perhaps unconscious, will occur primarily because if one has such a feeling, others will have a reciprocal feeling.

I must qualify this somewhat optimistic conclusion by noting that here on Earth, it is doubtful that many take the Greatest Happiness Principle literally. I am sure most people think they have greater obligations to their children than to others, and (perhaps) to their countrymen than to foreigners (Mackie 1977). Biology can give fairly ready explanations of this sense of variable obligation (in terms of kin selection and the like), and presumably what can be done for terrestrial morality can likewise be done for extraterrestrial morality (Ruse 1984).

The Categorical Imperative seems to be a philosophical version of the biological notion of reciprocal altruism simply made explicit (Murphy 1982). Reciprocal altruism urges you to care about others and to think of their well-being. What is this but to treat others as ends and not simply to use them? Of course, the Categorical Imperative in itself does not look at underlying causes of why you should feel the urge to obey it. In fact, as noted, Kant himself did not think that you should ever do anything for hope of conscious return. But, as we have seen, natural selection does not demand conscious awareness of your motives. In fact, in morality you might be better off doing things because you think they are right. (In this sense also, ethics parallels epistemology. The Darwinian is Humean in thinking that morality does not correspond to an external law. As noted, I would rather not talk of 'deception' here.)

What I argue, therefore, is that if, indeed, natural selection is at work on our extraterrestrials, we might expect to find something akin to the Categorical Imperative having evolved amongst them. Indeed, I would go further than this. My own hunch is that a lot of our moral tensions arise because evolution is not desperately concerned with solving every last moral problem. It gives us a rather broad and crude set of moral norms, which includes both the Greatest Happiness Principle and the Categorical Imperative. Usually these norms overlap; but sometimes they come apart a little. We are better off having both, despite the occasional tensions which arise. Perhaps therefore, our extraterrestrials will have moral philosophers, just as we do. Perhaps, as ours do, their philosophers will agonize over whether or not one should be a utilitarian, or whether one should be a Kantian, or perhaps something else.[17]

SEXUAL MORALITY

What about sexual morality? Much human moral thought centres on sexual desires and activities. Will there be a sexual morality for our extraterrestrials? Is rape wrong on Andromeda?

An affirmative answer presupposes that our extraterrestrials will be sexual beings. But, as I have pointed out, our present understanding of the evolution and maintenance of sex is not yet strong enough to predict with any reliability that extraterrestrials will necessarily be sexual beings. However, I would hazard a guess that they will be. Although sex certainly did not appear in order to bring about complexity and intelligence – natural selection is not that far-sighted – it is not easy to see how, without any sexuality, great complexity and intelligence could have occurred. Sexuality means good variations can be gathered together in one organism, much more quickly than could be achieved otherwise. Hence, I predict male and female extraterrestrials. (There is not much point in a third sex, although there could be sterile castes.)

But this does not mean we would have boy ETs and girl ETs, just as on Earth. Everyone might be hermaphroditic. Or there might be long phases of asexual reproduction, with very rare cases of sexuality. Perhaps indeed the extraterrestrials would revert entirely to asexuality. Or the mates might be minute, as in barnacles. Possibly the intelligent extraterrestrials would be all female, with little wart-like objects on their skin, these being males – no more than sperm sacs and penises (see Darwin 1851, 1854). If this were the way of extraterrestrial sexuality, then there would be little need of sexual morality.

But what if the denizens of our hypothetical planet have sexes of comparable size and intelligence, as with humans and other mammals? What place then for sexual morality? Let us take rape as a test case. I realize there are those who argue that rape is not primarily a sexual act, but really some kind of power trip (Brownmiller 1975). If it is simply this, then I presume that it is covered by the Greatest Happiness Principle or the Categorical Imperative. But I would argue – as I am sure most others would – that whatever else may be involved in rape (and probably power is), sex comes in at some point. Without sexuality, the very concept of rape collapses. Moreover – and this is important – almost universally rape is considered immoral. This holds, whether we are talking about eighteenth-century India, twentieth-century Africa, or twentieth-century North America. There are times when rape is not accompanied by much fear of retribution, such as in times of war. But rape is not considered morally acceptable. Rape is wrong.

Nevertheless, we cannot automatically assume that our extraterrestrials would think rape immoral. Why? Because, although the immorality of rape is a human constant, we cannot thereby assume that it will be a constant for other organisms, including extraterrestrial intelligent organisms. Certainly, if we look elsewhere in the animal world, we see that acts which look very much like rape, occur on a regular basis. For instance, among mallards, males will very frequently dash in and copulate violently with a female when she has already paired with another male (Barash 1977). Similar behaviour occurs in the mammalian world (Wilson 1975a, Dawkins 1976). Furthermore, there are good biological reasons why this sort of behaviour frequently occurs. If a male animal is prepared to attempt rape on occasion, then he is more likely to reproduce than otherwise.

But, you might object: because rape may be frequent in a species, it does not make it moral. In any case, the evidence we have from Earth, from the only extant intelligent species, is that it does not have to be *very* frequent for it to be considered immoral. This is true, but one can certainly think of situations where it could be quite frequent, despite intelligence. Suppose the extraterrestrial females all came into heat, as do the females of most mammalian species. If the males simply could not stop themselves from rushing in and copulating, there would be a lot of rape. And it could hardly be considered particularly immoral, since no one would have much choice in the matter.[18]

Furthermore, consider one of the major biological reasons why we humans think rape wrong – especially why males think rape wrong. Because humans take so long to mature, males co-operate in child-rearing, unlike most other mammals (Lovejoy 1981). Hence there are good reasons why human morality is trans-sexual, and why nobody (especially no male) wants some third party leaving his seed around in fertile places. If you have got to spend years raising a child, biologically there are reasons why you prefer it to be your own. However, if extraterrestrial females did all of the child-rearing, unaided, there might be simple moral emptiness when it came to rape. This is not to say the females would not have strategies to mate with the 'best' males (Hrdy 1981); they would![19]   (I am not saying we think rape is wrong simply because males might have to raise the children of others. I am looking for biological reasons why we might feel so strongly about it. Why is non-physically injurious rape put on a par with assault, or murder even? 'Because people get upset' is the whole point!)

THE MORAL CODE OF THE EXTRATERRESTRIAL

To sum up: extraterrestrial ethics will follow extraterrestrial epistemology. At the surface and even intermediate level, we shall find significant differences beween extraterrestrials and humans. In fact, the differences may be so great that it is impossible for us to find any common ground. We would simply be dealing with chalk and cheese. If what I have just been saying about sexual morality is correct, then, as with epistemology, these differences might run a lot deeper than most people imagine. We all know of differences between humans, but our common humanity also makes for many similarities. We may differ on meat-eating. We agree on rape. Outside our species the differences multiply, and we may not even find agreement about the immorality of rape.

Nevertheless, at the most basic level, there could well be certain fundamental similarities in morals between extra-terrestrials and humans, (that is assuming some kind of social life occurs, but if, like many male mammals, the males lead solitary lives, even this might not be true of males most of the time). However, perhaps naïvely, I suspect that there is probably either a cause or an effect relationship between intelligence and sociality. Certainly, in our world the two seem to be linked (Lovejoy 1981).

As in the case of epistemology, I have to admit that we are dealing with the products of evolution, and I, myself, am a product of evolution. Could there be a morality quite beyond my comprehension? All I can say is that, as with mathematics and science, I cannot conceive of what a non-human morality, at the most basic level, would be like. Could one have a morality which does not value happiness or which does not entreat one to treat one's fellow species' members as ends in themselves? I cannot conceive of what it would be like. As before, this has no bearing on the likelihood of such a morality. But it is certainly not something in which it is reasonable to believe.

HANDS ACROSS THE UNIVERSE?

Finally, what about the question of how extraterrestrials would behave towards us, and how we should behave towards them? The trouble is that neither our sense of morality nor their sense of morality has evolved to deal directly with situations like this. Therefore, in any ultimate sense, I do not think there are any definite answers.

But this is no real cause for concern. I do not know that our sense of morality has really evolved to give us any direct answers about our behaviour towards animals. This is surely reflected by the fact that people have different opinions about the proper way to behave towards animals. And this difference is not simply a question of not having enough information. Some people's intuitions lead them one way, other people's lead them another. My intuitions, for instance, lead me to believe that it is perfectly acceptable to kill animals in order to eat them. Peter Singer, who is a no less rational person than I, has intuitions leading him to believe that it is immoral to kill animals in order to eat them (Singer 1975).

Many of the problems arise because our sense of morality has really only evolved in order to enable us to deal with fellow humans. Indeed, I am not even really sure that it has evolved to give us a sense of how we should behave towards all humans. We certainly pay lip service to our need to respect members of remote and foreign cultures as ends in themselves; but, whether we really truly believe that we have a moral obligation to them, is, I think, a very debatable point, as more than one philosopher has pointed out in the past. Remember the qualifications I made earlier when talking of the Greatest Happiness Principle (Mackie 1977, Ruse 1984).

However, although we do not have a well-focused sense of morality about animals, we are not completely in the dark. It is not irrational for Singer to try to talk me over, or vice versa. We work by analogy from our feelings about our fellow humans. Much depends on the extent to which we think animals are like us and, also, on the extent to which we relate to animals and they relate back to us. For instance, we tend to treat as moral beings those animals with which we have a close warm relationship, and which respond. We consider their happiness as well as our own, and consider them as ends, not simply as means for us. For instance, I would certainly not think of hurting or eating my family dog. I care about his happiness, for his sake.

On the other hand, animals with which I do not have any particular relationship, or animals that do not strike me as particularly intelligent or responsive, receive far less moral concern from me. Although I do not think I would behave in a cruel manner towards pigs, I have a few qualms about killing and eating them. I do not relate to pigs as I do to dogs. Moreover, I do not think they suffer *Angst* at the thought of being killed. So for this reason also I do not think I have the obligations to pigs that I have to humans. (I am referring now to pigs' overall lives, and not necessarily to the cruelty which is practised in many modern slaughterhouses.)[20]

In short, what I am suggesting is that, inasmuch as other animals are like humans, and relate to humans in a reciprocal way, then human morality extends to them. And, inasmuch as animals are not like humans, human morality fades away. Certainly, when dealing with animals which do not respond in a friendly way to me, I do not feel a particular urge to respond in a friendly way to them. I do not, for instance, feel any sense of morality towards snakes; particularly towards those poisonous snakes which would kill me, if it suited their purposes. Nor do I see why I should treat snakes as moral beings. (I would accept qualifications about being wantonly cruel to snakes or gratuitously making them extinct. But why do we preserve such species – for their happiness, or for ours?)

Extraterrestrials merit conclusions analogous to those we draw about non-human terrestrials. Inasmuch as we can relate to them, and they seem to be like us, then no doubt our sense of morality will extend to them. Inasmuch as there is a failure of empathy, then our moral feelings will be that much diminished. If, in some way, we can relate to and communicate with the extraterrestrials and they with us, then presumably they will be that much more inclined to treat us as they would their fellow extraterrestrials. The bottom line is their (our) attitude and behaviour towards us (them). If they (we) are unbrokenly hostile, or if they (we) seem as though they (we) are going to manipulate us (them) for their (our) ends, then we (they) are going to feel no strong sense of moral obligation towards them (us). Nor indeed, is it easy to see why we (they) should feel any sense of obligation.

To illustrate the point I am making let us pick up where we came into this essay. In the movie ET, the little boy Eliott, who met the extraterrestrial, clearly felt a sense of warmth towards ET, and had moral sentiments about ET's welfare. Why was this? Simply because ET, in return, showed friendship and kindliness towards Eliott. And we would agree that it is appropriate to use moral language at this point. Eliott's moral feelings towards ET were not misplaced, nor were ET's moral feelings towards Eliott. It was right for Eliott to help ET and vice versa.

However, when we turn to less friendly extraterrestrials, like

Darth Vader, the villain in *Star Wars*, it is hard to see why anyone should have any sense of moral obligation to him. No one likes him very much, even his allies, and certainly the heroes of *Star Wars* did not feel any obligations to Darth Vader. Rather they tried to do him down at any and every opportunity. Nor were they wrong in so trying. Darth Vader simply wanted to do harm to others; consequently, in return, others felt no obligation towards him.

I would argue that the heroes, in fact, had no real obligation to Vader. He was outside the moral realm, at least as far as Luke Skywalker and friends were concerned. It is interesting to note that Vader's moral reclamation, occurring at the end of *Return of the Jedi*, came only after Luke saw Vader as his father, recognized a fellow (albeit fallen) Jedi, and showed love towards him. Vader then, in turn, responded to this love. Luke was his son! A neat case of kin selection at work, possible only when Luke and Vader saw analogous beings in each other. Confirmation of my whole line of argument![21]

So, in answer to the question of whether or not we have any moral obligations to extraterrestrials, and whether they have moral obligations to us, all we can say is that 'it all depends'. Are they enough like us that any kind of moral discourse is possible? Is their nature such that they could respond in any kind of friendly way towards us? Is our nature such that it is possible for us to respond in any kind of friendly way to them? There is hardly any point in talking about morality, if, for some reason, the extraterrestrials fill us with such a sense of loathing and disgust and fear that we simply cannot relate to them – and to do so would be deleterious to us. If the possibility of some sort of reciprocal altruism is there, then I see morality emerging; otherwise it may not.

As in the case of terrestrial non-humans, I am trying hard not to slip into some sort of naturalistic fallacy, arguing that morality is simply a function of evolution and that which has evolved is that which is good. I am not endorsing feelings of warmth or hatred, simply because they evolved. I am rather pointing to the fact that morality depends on some kind of moral sense. If a moral sense has not evolved, then morality is simply not possible. Moreover, if moral senses are far out of focus or otherwise blocked, say by mutual animal repulsion or hostility, then morality simply is not going to be able to function. There is no possibility of a moral relationship with extraterrestrials, unless in some way we can both get on the same plane. Perhaps such a plane will exist; as an evolutionist, I do not think it impossible.

CONCLUSION

Exploring the possibility of life elsewhere in the universe is full of philosophical interest. Such exploration puts in a bright light our own powers and limitations. By speculating on what other forms of life would be, we see more clearly the nature and extent of our own knowledge. This is true both in the area of epistemology and in the area of ethics. Such fairy-story telling does not prove anything empirical that we do not already know, but it does force us to think again about ourselves from a novel perspective.

In particular, we see a subjective or relative area in both epistemology and ethics. Much that we claim to know, both about the world and about ourselves and our moral relations to others, is very much a function of the kinds of beings we are. Specifically, it is a function of the kinds of animals that we are. We are not possessed of some kind of objective searchlight which enables us to view unfiltered reality. Rather, our vision of the world and our moral feelings for others are direct consequences of our evolutionary past, and there is no further justification.

Yet for all this, there is an underlying 'objective' universality to our knowledge, both in epistemology and in ethics. We may be mere animals, the products of natural selection, but our evolutionary past could not afford to let us believe absolutely anything. We have to be able to respond sensitively and productively to the world out there, and harmoniously with our fellows. The basic principles informing our thinking – principles to which I claim any intelligent (social) being would be sensitive – include such philosophical familiar friends as the central claims of mathematics (in epistemology) and the Greatest Happiness Principle (in ethics). I do not say these principles reflect ultimate reality. Frankly, I do not know what that would mean. I do say that their denial is inconceivable. And that is enough for me.

If the main themes of this essay are correct, that is enough for the inhabitants of Andromeda, also. But, for once, I doubt if anyone is going to show me to be wrong!

NOTES

1 By the time you read this, *Return of the Jedi* may be the biggest money-maker. Extraterrestrials abound!

2 For Aristotle, see *On the Generation of Animals*, 561b and *On the Movement of Animals* 669b 19. For Kant, see *Cosmologische Briefe* (1761), especially letters 6, 8, and 9.

3 To the second and succeeding editions of his work Whewell added an ever-growing introduction, responding to critics. His true fears appear even more blatantly in this introduction than in the formal text. In fact, Whewell wrote and had printed an earlier version of the *Plurality*, which included later-excised chapters making his theological motivations more obvious. See Ruse (1975a).

4 In a critique of Puccetti (1968), McMullin (1980) makes this point very well. Incidentally, these two contributions to the extraterrestrial life debate show very well that the theological underpinnings of the debate are still with many of us.

5 For a philosopher, I intend to plunge fairly deeply into scientific waters. I will try not to presuppose technical knowledge, but a grasp of modern thought about evolution would help. As background reading I recommend Dobzhansky *et al* (1977); Ayala and Valentine (1979); and Futuyma (1979). I give a thoroughly opinionated survey in Ruse (1982c). The September 1978 issue of *Scientific American* contains absolutely first-class articles on evolutionary thinking today.

6 Before Darwin, the natural creation of new life – 'spontaneous generation' – was a much discussed topic. However, the canny author of the *Origin of Species* (1859) realized full well that any discussion of the topic would be bound to be incomplete and thus troublesome for his theory. Hence, wisely, Darwin said nothing at all about life's initial creation. And for nearly one hundred years, evolutionists took his example to heart and likewise said nothing. Only recently, as progress on the coming of life has occurred, has it reappeared in evolutionary discussions. See Farley (1977) for more details.

7 See Popper (1974) for the claim that selection is a tautology, and Ruse (1981a) for the denial.

8 For a fascinating discussion of how selection can lead to different strategies for survival, see Oster and Wilson (1978).

9 From here on, I am going to be assuming the general correctness in approach of modern human 'sociobiology' (as I have done in Ruse 1979b, 1981c, 1982c), the neo-Darwinian thesis that human behavioural and intellectual traits and capacities are in major

part direct functions of evolution through selection. This was Darwin's own claim in the *Descent of Man* (1871), and it has been taken up recently by such thinkers as E.O. Wilson (1975a, 1978); Lumsden and Wilson (1981, 1983); Alexander (1979); Symons (1979); Hrdy (1981); van den Berghe (1979); and Barash (1977). I must warn, however, that there are strong and strident critics, for instance Sahlins (1976); and Allen *et al.* (1977). If the extreme critics are right, and if indeed nothing human (other than our anatomy) has a link to biology, then I just do not know that you could make any claims about extraterrestrial knowledge. But, as I argue in Ruse (1982b, 1982c), I do not think the critics are right. My arguments in this essay, however, do not depend on some of the more specific and controversial claims by sociobiologists (e.g. Lumsden and Wilson's gene/culture co-evolutionary model).

10 The literature on this subject is so vast that I really do not know where to begin with the references. The major pertinent recent works are probably Griffin (1976), and Crook (1980). A useful short discussion, with many references, is Crook (1983). As you will see in a moment, I rather dodge the whole issue by claiming it is not important (to me)!

11 Indeed, as Kuhn (1962) has truly pointed out, scientists within the same culture frequently find it impossible to communicate, one with another. As will become clear, I do not subscribe to Kuhn's radical position.

12 Putnam invites us to consider the possibility that we are all 'brains in a vat', that is, disembodied brains, supplied with appropriate nutrients and electrical inputs to make us think that we are normal human beings – seeing, feeling, loving, laughing, and so forth. Such brains have their reality. They 'know' chairs and tables and each other. They are 'internal realists'. But what of the world outside the vat – 'metaphysical reality'? Putnam argues that such an outside world is an incoherent notion, judged from our human vantage point. Coming from a biological perspective, a perspective about which Putnam has some rather odd ideas, and not having read his work before completing the first draft of this essay, I am nevertheless encouraged to find that I make the same distinctions of reality as he, and that I endorse common-sense (internal) reality, whilst feeling most uncomfortable about ultimate (metaphysical) reality. But we are both Kantians of a kind, so perhaps the consilience was to be expected.

13 This is the famous 'Molyneux problem'. Can a man, blind from birth, able to distinguish cube from sphere by touch, be able to tell apart cube and sphere as soon as he is given sight? The answer is 'no'. See Berkeley (1963) and especially C.M. Turbayne's introduction.

14 Causation I take to be governed by kinds of 'epigenetic rules' as

discussed by Lumsden and Wilson (1981) – genetically caused traits which affect and govern the way we think. I take them to be akin to the 'regulative principles' of modern Kantians. See Ruse (1982b) and Körner (1960).

15 Like all would-be Kantians, I am haunted by the Ding an Sich problem – how can one know about ultimate reality? As explained previously, I think that if there is an ultimate reality, a lot of the problems about it are bogus. It does not exist, as it were, like the inside of an orange without the skin (*i.e.* the external sensations). The inside of an orange is still an object of touch, smell, taste, and so forth. Reality, with sensations stripped away, is not. In fact, 'reality with sensations stripped away' does not make too much sense to me. Putnam (1981) argues that ultimate (metaphysical) reality is really an incoherent notion. Perhaps it is simplest just to leave matters at that.

16 It has always seemed to me that, in their desire to hammer E.O. Wilson for being sexist and fascist, his critics miss entirely the extent to which his (and other sociobiologists') vision of humankind is one of a species *sharing* traits. The New York executive and the Kalahari bushman do what they do for the same reasons. See Ruse (1979b) and (1982c) for more on this theme.

17 I am not suggesting that all moral disagreements come from clashes of ultimate principles. I am sure that much of the disagreement over (say) censorship comes because we just do not know if pornography corrupts. But, I do suggest that some disagreements come simply because ethics 'break down'. Making an example of a drunk driver has clear utilitarian value – yet, as a Kantian, it just does not seem to me very fair. Philosophers make a living from arguing their way out of these moral swamps. I wonder if there really is a way out. As an evolutionist rather than a Platonist or Christian, and knowing that the products of evolution tend to be a bit rough and ready, I am not sure there has to be.

18 One might argue that, by definition, rape occurs only when there is choice on the part of the man, and hence is necessarily wrong. I find this unimpressive. I am talking of forced sexual intercourse, which the male would enjoy, the female would dislike, and which others would not want to see happen. That is close enough to 'rape' for me.

19 Again, let me reassure the reader who fears that sociobiology is all sexist claptrap, that no one is saying that females are defenceless. Unless a female's reproductive strategy is as good as that of a male, she might just as well be one. If a male has (on average) more offspring than a female, then at once there will be strong selective pressures towards giving birth to males. Before long, the rarity of females will make it advantageous to have female offspring, and so the sex-birth ratios will balance

out. Sociobiology preaches the biological equality of the sexes.

20 Some readers may think I am coming awfully close to the naturalistic fallacy here, arguing that because I do not feel friendly towards pigs, I have no moral obligations towards them. This is not so. I certainly do not think I should be wantonly cruel to pigs, but that is because such cruelty lessens me and other humans. It is not because I treat pigs as ends in themselves, or care about pig happiness for its own sake. My point is that you cannot have a moral relationship to something you do not recognize as a moral being. You cannot have a moral relationship with a table. What I am arguing is that moral feelings are a function of evolution, and evolution does not suggest that humans have developed moral feelings for pigs. Indeed, today, evolution emphasizes that selection works ultimately for the individual, not for the group. Hence, intra-specific feelings (e.g. those that we humans have for our fellows) have to be related to the individual. This is my position. Animals become moral beings only insofar as we can extend human-like feelings towards them. I love my dog like a son – literally! See Williams (1966); Dawkins (1976); Ruse (1981c).

21 Note also the neat resolution of the Han Solo/Luke Skywalker/Princess Leia triangle. In *The Empire Strikes Back* all sorts of tensions arise as Han and Leia draw together, apparently excluding Luke. But then it turns out in *Return of the Jedi* that Luke and Leia are siblings. Down comes the incest barrier! This is one of the big moral absolutes for sociobiologists, given the horrendous biological effects of close inbreeding (Alexander 1979, Wilson 1978, van den Berghe 1979). Obviously natural selection has been at work in George Lucas' world, just as I predicted it would!

Part IV

ULTIMATE QUESTIONS

Many people think – or behave as if they think – that Darwin became an atheist on a Friday morning and then, as a kind of encore, he became an evolutionist in the afternoon – or something like this. Actually, however, the truth was quite otherwise. It is true that in his early twenties, Darwin, who had been intended for the Church, lost his childhood faith in the literal truth of the Bible, with all of its attendant miracles. But as this theology (known technically as 'theism') went, another took its place. With the geologist Charles Lyell, not to mention family members like his older brother Erasmus, Darwin embraced the belief in God as unmoved mover, as one who can do everything through unbroken law (a position known technically as 'deism').

Darwin's new religion, one which lasted right through the genesis of the *Origin*, after which under the influence of T.H. Huxley he drifted into uneasy agnosticism, was very much a function of a society in the thrall of an industrial revolution. If the English can make cloth, not by hand but by machine, can God working by law, not miracle, do any less? At least, so thought Darwin, grandson of one of the most successful of industrialists, Josiah Wedgwood of pottery fame and (significant) fortune. Given that organic evolutionism is the epitome of a law-governed vision of the universe, one might more truly say that Darwin's *Origin* came about because of religion rather than despite it – and this is apart from the already-discussed influence of natural theology on Darwin's thinking about the design-like nature of organisms.

Of course, we all know that after the *Origin* there was a mighty clash between science and religion, epitomized by the notorious debate between Huxley and the Bishop of Oxford, Samuel Wilberforce, at the British Association for the Advancement of Science in 1860. Like many popular tales, however, this is only part of the truth. For all the pyrotechnics, acceptance of Darwin's ideas (especially about evolution *per se*) was remarkably rapid. That is, with the exception of the southern part of the United States, where a particularly virulent form of Protestant evangelical biblical literalism took hold – a grip that persists to this day. Erupting with monotonous regularity, the literalists (known as 'fundamentalists') had their most recent triumph in the passing of a law in 1981 in the state of Arkansas, mandating the teaching of so-called 'creation science' (otherwise known as the early chapters of Genesis taken absolutely literally) alongside Darwinism in the schools of the state.

Thanks to the vigilance of the American Civil Liberties Union, suit was brought and the law declared unconstitutional. I was myself a witness, testifying to the non-scientific nature of creationism. As it happens, my stance has been much criticized by my fellow philosophers, for the core of my position was that there is a fundamental difference (of logical type) between the *scientific* claims of the Darwinian and the *religious* claims of the creationist. Many feel uncomfortable with a position which presupposes such a stark 'criterion of demarcation', arguing to the contrary that most products of the human psyche exhibit the same defining features. Science and religion have much in common.

I have not repented of my belief that science and religion are different – Darwinism and creationism specifically so (Ruse 1988b). Nevertheless, as you will realize, I see now that there is significant overlap between the scopes of science and religion, especially with respect to moral dictates. Moreover, whereas once I thought that science and religion properly understood speak to different spheres of human experience (whether truly or not) and thus cannot come into conflict, now I am no longer so sure. I certainly have no wish to jettison all Christian insights, and indeed (immodestly) would like to think that the single essay in this section, hitherto unpublished and the final one of the collection, shows a certain subtle sympathy for Christian morality. Nevertheless, if my argument about the foundational moral nihilism of Darwinism holds true, then Western religion as we know it is under grave threat indeed. The science/religion relationship is headed once again for trouble.

Ultimately, this will be for others to judge. I do note wryly that 'unpublished' implies 'untested'. Here, I will simply use my enthusiasm to confirm how for me the Darwinian paradigm is a never-ending source of challenge and excitement. And that is a good note on which to end.

Chapter Ten

EVOLUTIONARY THEORY AND CHRISTIAN ETHICS

Are they in harmony

Thanks to the unflagging efforts of the creationists, the oft-times troubled relationship between science and religion has again been brought to the fore. It is argued that you cannot subscribe to the modern theory of evolution through natural selection, and at the same time be a sincere practising Christian (Morris 1974, Gish 1973). However, the opinion of reflective thinkers, in both science and religion, is that this charge is quite mistaken (Montagu 1984). To quote the nineteenth-century philosopher/scientist John F.W. Herschel: 'Truth cannot be opposed to truth.' If God created through an evolutionary process, then this is for practising Christians to recognize and appreciate, not deny (Ruse 1979a, Durant 1985).

As one who has laboured long in the field of evolutionism, I do not at all disagree with the spirit of this general conclusion. I would regard it as a blasphemous denial of God-given powers of sense and reason, to pretend that salvation requires the taking literally of ancient Jewish myths about beginnings. Yet, as one who is concerned to push our thinking up to and beyond present limits – especially in the field of social behaviour ('sociobiology') – I fear that there are again growing problems, stemming from evolutionism, standing in the way of the would-be believer. In particular, I suspect that hitherto-unappreciated implications of evolutionism for ethics cast grave doubts on certain claims the Christian is obligated to make.

I begin with some scientific background, moving then to ethics (morality) and to possible tensions with true religion.

HUMAN EVOLUTION

Human beings are animals, and as such are the end product of a long, natural process of evolution, which began some three and a half billion years ago. This process is primarily that first fully articulated by Charles Darwin in his *Origin of Species*, in 1859. More organisms are born than can possibly survive and reproduce, and there is consequently an ongoing 'struggle' for limited resources and mates. Because organisms vary for various reasons, the survivors (the 'fit') tend to have features not possessed by the losers. There is consequently a 'natural selection', which, given enough time, results in full-blown evolution. As Darwin and his successors have always emphasized, the evolved features of organisms work or function. They are 'adaptations', helping their possessors in life's struggles. Darwin's selective mechanism, therefore, is the naturalistic answer to the designing god of Archdeacon Paley (Ruse 1979a, 1982b).

What evolution demands is that there be an ever-replenishable supply of new variation, and that successful features be preserved down through the generations. Darwin himself had little idea about the underlying causes of heredity; but now, in this century, we have a full-blown theory. It is realized that the units of heredity, the 'genes', are generally transmitted unchanged from generation to generation; that occasionally, for natural reasons, the genes do alter (they 'mutate'); that (as Darwin always emphasized) there are no non-natural teleological forces guiding change; and thanks to the work of molecular biologists, that ultimately everything relates back to the molecules. This is not some sinister anti-religious materialism; but, simply, the recognition that modern science has neither place nor need for mysterious life forces like 'entelechies' (Ayala 1974).

Important here is the fact that, in essential respects, humans are typical products of evolution. Of course, we are unique. But, then, so is *Drosophila melanogaster* (a species of fruit-fly). The point is that, considered biologically, we humans work in much the same way that every other organism does. Our physical features are controlled by the genes, within our bodily cells, and all the evidence is that they result from the pressures of natural selection, as is true of other animals and plants. We now know that it was a mere six or seven million years ago that our ancestors broke from the great apes. Even today, at the biochemical level, we are more closely related to the chimpanzees than the chimpanzees are to the gorillas (Pilbeam 1984).

These facts are not truisms; nevertheless, although probably most people do not recognize the closeness of our links with the brutes, they embody claims that are not particularly controversial. But most people would still argue for human distinctiveness on the grounds of our culture or our social behaviour. Where is the baboon Shakespeare or the chimpanzee Mozart? Or, for that matter, where is the gorilla Hitler?

That humans do have a distinctive culture is incontrovertible. Nor would anyone, least of all me, claim that this culture is *purely* a result of the genes – anymore than is height, weight, or health. However, the evidence is strong that the genes, as promoted by natural selection, do have a significant causal input into human social behaviour, and consequently into culture. This is a fact with massive empirical backing. To the Darwinian, it is hardly a surprise, for clearly an organism's behaviour is as crucial to its reproductive well-being as is any physical feature. Humans need their eyes and mouths and feet and digestive systems to survive successfully. Why should their behaviour be an exception?

The exact ways in which genes control behaviour is still under intensive investigation (Lumsden and Wilson 1981, Boyd and Richerson 1985, Cavalli-Sforza and Feldman 1981). Most revealing is what happens when things go wrong. Medical researchers have compiled an increasingly long list of quite bizarre behaviours, due to relatively minor changes down at the level of the units of heredity. Lesh-Nyman's syndrome, for instance, is a genetically caused ailment leading to compulsive self-mutilation. The unfortunate afflicted have to be physically constrained all their lives (Hilton *et al.* 1973). However, their tragedy is science's gain, for through such aberrations we are able to trace the usually beneficial behavioural effects resulting from our basic biology.

I shall not labour the point. If we were not *Homo sapiens*, and thus with a peculiar interest in this species' nature, there would be far less reason than is presently shown for picking out our species for special attention. Humans were produced by the natural selection of non-directed variations, and they reveal this legacy in their behaviour as well as in their physical attributes.

SEXUAL MORALITY

Already, even with but the bare bones of our scientific position laid out, it is possible to make some comments about morality. In particular – and not surprisingly, given the central role that reproduction plays in the Darwinian evolutionary process – one can make comments about some of the sexual prescriptions and prohibitions that have been made in the name of Christianity.

I shall go right to the heart of the matter, looking at the Christian position on the value of sex in itself. Jesus, in fact, had comparatively little to say on the matter (although there is the famous comment in the Sermon on the Mount about lusting and adultery in one's heart), but St Paul certainly repaired the omission. 'It is good for a man not to touch a woman' (1 Corinthians 7, 6). It is undoubtedly true that many sincere Christians have felt that the apostle's views are somewhat less than binding. Nevertheless, the attitude is rather negative: 'it is better to marry than to burn' (1 Corinthians 7, 9), and undoubtedly the views have had great influence. One has only to think of the Catholic Church's attitude towards sexuality, even today. If sex is such a good thing, why is the Pope not married?

As an evolutionist, looking at a morality infused by this attitude towards sexuality, one must tread carefully. I recognize that you cannot deduce moral dictates ('Sex is good and should be cherished') from factual claims ('Sex is important in evolution'). To assume otherwise is illicitly to cross the 'is/ought' barrier: to commit what G.E. Moore (1903) called the 'naturalistic fallacy'. For this reason, whatever the facts, whether sex be essential or irrelevant to the evolutionary process, one can still mount an argument in favour of chastity or continence of some form or another. Nevertheless, assuming that a loving God does exist, I think it improbable that He would have put in motion an evolutionary process that centres so directly on sexuality, and then, in some way, have declared the sexually active life as less than ideal. God may well make demands on us as Christians, for instance to behave lovingly, within and without sexual encounters. I shall speak more of these matters shortly. The point is that, to the evolutionist, recognizing the central place of reproduction, a priori declarations about the inherent worth of sexual abstention ring false.

Turning now to more specific sexual dictates, let me take the norm which, although widely ignored in North America, still has major effects in the world at large. I refer to the Catholic prohibition against birth control, reaffirmed in 1968 in the encyclical *Humanae Vitae*, by Pope Paul VI.

> It is in fact justly observed that a conjugal act imposed upon one's partner without regard for his or her condition and lawful desires is not a true act of love, and therefore denies an exigency of right moral order in the relationships between husband and wife. Hence, one who reflects well must also recognize that a reciprocal act of love, which jeopardizes the responsibility to transmit life which God the Creator, according to particular laws, inserted therein is in contradiction with the design constitutive of marriage, and with the will of the Author of life. To use this divine gift destroying, even if only partially, its meaning and its purpose is to contradict the nature both of man and of woman and of their most intimate relationship, and therefore, it is to contradict also the plan of God and His will. (Reprinted in Baker and Elliston 1984, p. 173)

Again, one must tread carefully. Note that the late Pope's condemnation of artificial means of birth control does not stem solely from the way that we are. It stems from a judgement of what is 'natural' for human beings, combined with the claims that this was God's plan or design, and that this is what He wants us to follow. Given these additional premises, then, if God wants us to stand on our heads during sex, this is final, no matter what the facts of evolution may be.

However, I would challenge the claim that the immorality of artificial birth control is (in part) a consequence of its unnaturalness. At least, I would challenge the claim that the only true biological purpose of sexual intercourse is to lay open the possibility of conception at any time (or even, only during the so-called 'fertile periods'), and that consequently without such possibility we have entered the realm of the unnatural. To argue in this way is to show a gross misconception of the nature of the evolutionary process. What must be recognized is that natural selection is opportunistic. What works, succeeds. If some feature can serve a new purpose, or two purposes at once, then so be it (Futuyma 1979). If God works through the medium of natural selection, then you can no longer say that He had one and only one purpose in mind for any particular characteristic.

Undoubtedly, sexual intercourse, involving pleasurable sensations for both male and female, evolved for the direct purpose of fertilization. But, this is not to deny that it may also have taken on other biological virtues, in their way equally valid – equally 'natural' (Symons 1979). The bird's forelimbs started as fish's fins. Is flying unnatural in a sense that swimming is not? Specifically: lungs evolved for breathing, and they are still used in this way. Is talking unnatural (and, hence, immoral)?

In the case of human sexual intercourse all the evidence is that, direct reproduction apart, it plays a vital role in pair-bonding, something absolutely crucial in *Homo sapiens*, where the young require so much attention, preferably from both parents (Lovejoy 1981). Indeed, it is highly plausible that many of the more pleasurable aspects of intercourse stem from this factor, alone. If fertilization were all that counted, we could couple fleetingly, like other mammals.

In other words, sex for sex's sake is quite natural, and if God stands behind the evolutionary process, then according to Catholic theology He ought to cherish such activity – perhaps, with the proviso that it occur within the bonds of ongoing relationships. And if, through our technology we have made less pressing the need for continuous fertilization (because, for example, of improved child-care), the case for the moral virtues of the pill or of condoms is made complete. (I emphasize that, as an evolutionist, I am not necessarily a sexual radical, preaching the virtues of promiscuity. Sex, like all human activity, is subject to general laws of morality: see also Ruse, 1988a.)

Finally, let me turn to the highly contentious issue of homosexuality. Historically speaking, it turns out that in practice the Church has had a variable attitude towards homosexual activity, and, today, even conservative churchmen generally realize that homosexual orientation is not something over which people have much control (Boswell 1980). Nevertheless, homosexuals have frequently suffered in the name of the Lord, and many Christians continue to condemn homosexual behaviour. Again, St Paul is the authority.

> For this cause God gave them up unto vile affections: for
> even their women did change the natural use into that
> which is against nature: And likewise also the men,
> leaving the natural use of the woman, burned in their lust
> one toward another. (Romans 1, 26-7)

This position did not originate with St Paul. It goes back to the Holiness Code of Leviticus, and indeed can be found in Plato (in the *Laws*). Coming towards the present, Aquinas was succint and unambiguous. He thought homosexuality was worse even than rape. The latter is just a violation of a human being. The former violates God.

> The developed plan of living, according to reason, comes
> from man; the plan of nature comes from God, and
> therefore a violation of this plan, as by our natural sins, is
> an affront to God, the ordainer of nature. (Aquinas,
> *Summa Theologica* 2 ₂2ae, 154: 12)

But the question remains! Is homosexuality biologically unnatural? Modern evolutionary theory suggests that this claim is highly questionable. Certainly, we can say with some confidence that homosexual activity is not (as everyone from Plato on down seems to have assumed) a phenomenon exclusively restricted to humans. Indeed, it is no exaggeration to say that *every* animal species studied with care shows some such behaviour (Weinrich 1982).

Moreover, there are plausible theoretical reasons as to why such behaviour might have evolved (referring now to human behaviour, in particular). For instance, one hypothesis focuses on the so-called mechanism of 'kin selection'. Ultimately, the only thing that counts in evolution is transmitting your genes, or rather *copies* of your genes. But close relatives share copies of the same genes. Hence, inasmuch as one's relatives reproduce, one is reproduced oneself, vicariously, as it were. Thus natural selection can (and does) promote features whereby relatives are led to aid each other, in the cause of reproduction. In this context, it has been suggested (with significant empirical backing) that homosexual orientation might be part of a (selectively caused) reproductive strategy, whereby individuals are turned away from their own direct ends, thus being freed to aid relatives in various ways. Because their brothers and sisters and cousins and others reproduce, the homosexuals 'reproduce' (Ruse 1981f, 1988).

I do not say that this, or related suggestions, is necessarily correct. But we are clearly at the point when only the ignorant can claim confidently that homosexuality (orientation or behaviour) is unnatural. And this being so, the major ground for Christian (and Jewish) condemnation of such sexuality crumbles. As in the case of birth control, I recognize that the Church has never opposed homosexuality solely because it is unnatural. Prohibitions have been backed by claims about God's wishes; Aquinas makes this most clear. And, as before, whatever its biological status, one could, possibly, continue to condemn homosexuality. But the traditional arguments must be changed.

Three points are enough. Modern evolutionary theory clearly throws important light on many moral claims about sexuality that have been made in the name of Christianity. Although I suspect that many (particularly if they are more liberal thinkers) will find what has been said, thus far, neither surprising nor dreadfully upsetting. What they will feel, nevertheless, is that I am just skimming over or around the really important questions. True Christian morality deals with sexuality only incidentally, as part of the larger questions. Real Christian practice centres on the love commandment: 'Thou shalt love thy neighbour as thyself' (Matthew 22, 39).

Here, surely, we have the starkest of contrasts with everything the evolutionist holds dear. If evolution, supposedly, stems from natural selection brought about by a struggle for existence, then surely, ultimately, we are left with combat and selfishness. Evolution makes us tune in on our own ends, hostile and suspicious of everyone else. Christ's message, consequently, is one which is opposed to our brute nature, calling us to higher things. In fact, it is precisely because we are naturally self-serving, that Jesus came to preach His message. This was why His sacrifice was necessary. Here, then, the Christian agrees with the evolutionist that there is no morality to be derived from the study of human nature; but perhaps to the annoyance of the evolutionist, the Christian concludes, therefore, that this ends all interesting discussion (Trigg 1982).[1]

I believe that this conclusion is altogether too quick. I have agreed that you cannot use an understanding of human nature as a conventional foundation for moral dictates; but I would argue that there is more to the question than this. To make the case, I turn again to science.

EPIGENETIC RULES

I would like to pick up the thread with that which is distinctive about human beings, namely our (comparatively) large brains and the associated thinking power. There is little debate that this feature (or features) has come about conventionally, namely through the forces of natural selection making towards adaptive ends. The fossil record demonstrates clearly that the past four million years has seen a steady increase in brain size, and this has been accompanied by an equally steady increase in tool use and sophistication. That humans today succeed, in great measure because of their intelligences, hardly needs comment (Isaac 1983).

But let us think for a moment about how the human 'behavioural control box' works, and how it leads to and influences behaviour. In theory, there are a number of possible options. At one extreme, the brain and mind (if such there be associated with it) might rigidly predetermine all action (Ruse 1986a, 1987b). Everything is done purely automatically. Ants work in this way, and there is much to recommend their system. No time is wasted on decisions, and a comparatively simple mental apparatus is all that is needed. At the other extreme, the brain would be like a super-powerful computer, where every problem is weighed and assessed rationally. An animal takes no actions, until it has determined fully what would be the optimum path, in order to promote its own reproductive interests.

For fairly obvious reasons, neither of these options was particularly attractive to organisms like humans, and equally obviously we have taken neither. The trouble with being locked rigidly into behaviour is that one has no recourse when things go wrong. At the slightest environmental disruption, organisms begin to die wholesale. From the perspective of an ant parent, this does not matter desperately, for it can always produce lots more. From a human perspective, such rigidity is terrible. Each organism requires so much care that a parent just cannot risk losing it at the merest situational tremor.

The super-brain route is not much better. Each human would require more resources and care than it does now. And we would be forever making up our minds. Our situation would be rather like that of those early chess-playing computers, which surveyed every possible move, but which were totally useless. Their task was so great, they could never get to a decision.

In fact humans seem to have taken a middle course, and the new, much more successful generation of chess computers tells us about this course. Now, such computers are programmed to follow certain proven-successful strategies, reacting according to their human opponents' moves and abilities. The computers sometimes lose; but more often than not they win. Similarly with humans. Our minds are not *tabulae rasae*. Rather, they are structured according to various innate dispositions, which have proven their worth in the past struggles of proto-humans. These dispositions do not yield fully explicit, innate ideas (of the kind attacked by John Locke, in his *Essay Concerning Human Understanding*); but, as we grow, triggered and informed by life's experiences, the dispositions incline us to think and act in various tried and trustworthy patterns.

Such dispositions or propensities are known, technically, as 'epigenetic rules' (Lumsden and Wilson 1981). There is growing empirical evidence both as to their nature and of their widespread importance. One of the best-studied rules concerns fears and phobias. Clearly there is nothing absolutely rigid going on here. Certainly with respect to strangers, the actual content of fears and prejudices requires environmental input – else, why did the Nazis go to such lengths to indoctrinate their children against the Jews? One can see also precisely why an epigenetic rule (or rules) of this type would be cherished by selection. The protohuman who learnt at a very early age to avoid potentially dangerous things, inanimate or animate, nonhuman or human, was much better prepared for life's struggles than the protohuman who did not. I say this notwithstanding the fact that in modern technological society such a rule (or rules) might be dated. We would do better to learn to fear light sockets than snakes, and with our horrifying powers of destruction the fear of strangers is a mixed blessing.

THE EVOLUTION OF MORALITY

What has all of this to do with morality? Still staying with science, one of the most important findings of evolutionists – forecast by Darwin in the *Origin*, and confirmed fully in recent years – is that the best path to reproductive success is not necessarily one of bloody combat: 'nature red in tooth and claw'. More particularly, we get much further ahead by co-operating. A cake shared is far preferable to no cake at all, or even to the whole cake if the cost is serious personal injury. Given that we are all, at times, liable to be at a disadvantage – through youth, or illness, or plain bad luck – there is much to recommend the evolution of features and inclinations driving organisms to co-operate one with another (Trivers 1971).

Technically, this co-operation is known as 'altruism'. Although this is obviously a term derived metaphorically from the human world, for today's biologists it refers simply to features and behaviour involving effort on behalf of or to the benefit of others, ending in increased chances of personal reproduction. It does not, as such, imply or demand the conscious intention covered by (literal) altruism. 'Altruism' to altruism, has the same relation that the physicist's notion of 'work' has to what we call work.

Nevertheless, just as when we work – say, mowing the lawn – we do 'work', so also the claim is that (literal) altruism and (biological) 'altruism' are connected. In particular, it is argued that, in the case of humans, in order to make us perform 'altruistically', because we do indeed (for good biological reasons) have selfish feelings, we have laid over us (literal) altruistic inclinations. And, as you might imagine, given what we have just seen, the claim is that our altruistic dispositions are mediated through the epigenetic rules.

Note that it is a crucial part of the biological explanation of morality that it exists in order to get us away from the literally selfish or otherwise unpleasant motives that we might have from the other epigenetic rules. There is therefore no simple identification of the good with that which has evolved as mediated through the rules. I am *not* justifying the killing of Jews through a claim that the Nazis were motivated by a rule for prejudice and fear. Quite apart from the fact that the Holocaust clearly involved many non-biological factors, the whole point of morality is to act against other emotions, because humans have more ends than one. We must look after ourselves, but we must get on with others.

The position of the modern evolutionist, therefore, is that humans have an awareness of morality – a sense of right and wrong and a feeling of obligation to be thus governed – because such an awareness is of biological worth. Morality is a biological adaptation no less than are hands and feet and teeth (Mackie 1978, Murphy 1982, Ruse and Wilson 1985).

THE LOVE COMMANDMENT

What has all of this to do with Christianity? Let us move back to the love commandment: 'Thou shalt love thy neighbour as thyself.' If modern biology can yield us something like this, then we might be on the road towards a happy relationship between evolutionary theory and Christianity – otherwise not. As biblical scholars know full well, however, part of the difficulty one faces now is that the commandment appears in somewhat different settings – in the Old Testament and in the New, in the Gospels and in St Paul's writings – and there are various traditions within Christianity as to the commandment's proper interpretation and full import.

For brevity, I will content myself by responding to two fairly basic interpretations, which I will refer to as 'weaker' and 'stronger'. In the weaker understanding, as one interpreter has put it, 'neighbor love is identified with the core of natural morality' (Wallwork 1982, p. 302). One's obligations are to be a good family man or woman, to be decent and kind to one's friends and acquaintances, to 'render unto Caesar that which is Caesar's' (not forgetting the second half of this command), and to be prepared to lend a hand to a stranger in need (shades of the good Samaritan). There are no extreme expectations or obligations in this weak interpretation. One is certainly not expected to make a martyr of oneself with enemies.

As you might expect by now, this kind of morality causes no problems at all for the evolutionist. Indeed, it is basically just what he or she argues emerges from the selective process. As we have seen, altruism is put in place to promote 'altruism'. We are better off if we work together and co-operate than if we lead selfish, hostile, lonely existences. So we have evolved sentiments of friendliness and obligation – the very 'natural morality' of the weaker interpretation. We should help our neighbours, because they in turn will help us, and so we will all benefit. But, there is neither cause nor good evolutionary reason to make a (dead) fool out of yourself in the name of morality.

Yet can the evolutionist claim even this much fellow sympathy with the Christian? Even the weaker interpretation of the love commandment demands a genuine moral effort. You must love your neighbours because it is right to do so, not because you hope for some personal gains. There is nothing wrong with a straight business transaction – if you scratch my back, I'll scratch yours – but it is not morality. And this is all that the evolutionist seems to concede.

However, this objection is to miss the full force of the evolutionist's case. It is true that, causally, what is at work is a process aiming ultimately for individual reproductive benefit; but there is no implication that at the conscious level people will be scheming how best they can maximize personal benefits from any transaction. Rather, indeed, the opposite is the case. We are selfish brutes, it is true. But, laid on this is a genuine sense of morality. We do good because we think it is good. The evolutionist's case is that, precisely because we think the good is good, we function a lot better as co-operators, than if we were always looking for personal gain (Wilson 1978). And, in any case, no evolutionist thinks that for every kind act you expect immediate return. Morality is like an insurance scheme. You throw your policy into the general pool, and then can draw on it as needed.

I turn now to the stronger interpretation of the love commandment. Here, the need is to read the Sermon on the Mount (and related passages) rather literally.

The love commandment is considerably heightened beyond natural morality when theological attention shifts to the issues of human sinfulness and the need for redemption. Then, the commandment is commonly interpreted as demanding the purity of heart insisted upon in the Sermon on the Mount. As Rudolf Bultmann (1951, 1, pp. 13-18) observes, the meaning of the antitheses of the Sermon on the Mount – which are introduced with the words 'You have heard that it was said to the men of old. ... But I say to you ...!' – is this:

> What God forbids is not simply the overt acts of murder, adultery, and perjury, with which law can deal, but their antecedents: anger and name-calling, evil desire and insincerity (Mt. 5:21f.,27f.,33-37). What counts before God is not simply the substantial, verifiable deed that is done, but how a man is disposed, what his intent is ....*God demands the whole will of man* and knows no abatement in His demand ... What, positively, is the will of God? *The demand of love.* 'You shall love your neighbor as yourself!' ... The demand for love surpasses every legal demand; it knows no boundary or limit; it holds even in regard to one's enemy (Mt. 5:43-48). The question, 'How often must I forgive my brother when he sins against me? Is seven times enough?' is answered: 'I tell you: not seven times, but seventy times seven' (Mt. 18:21f.par.Bit.).

> Calvin ([1559] 1962, p. 354) further illustrates the radicalness of this second usage when he argues that we are totally condemned, whatever our intentions, if any 'thought be permitted to insinuate itself into our minds, and inflame them with a noxious concupiscence tending to our neighbor's loss.' (Wallwork 1982, pp. 302-3)

I am aware that today's students of the New Testament warn that passages such as these should not necessarily be taken at face value. The Sermon on the Mount, for instance, is probably not the verbatim report of a speech by Jesus, but something put together by one group of followers, radical Jewish thinkers facing hostility, in the early years of the Christian movement, from their fellow Jews (Betz 1985). 'Enemies' in this context refers to immediate opponents, and should not therefore (to move to a modern context) at once be generalized to include the Russians or Iranians. Nevertheless, many Christians have taken the love commandment in this strong way, and some still do – Quakers, for instance. So comparing it against the implications of evolutionary theory is more than a purely academic exercise.

I need hardly say that the modern biologist looks somewhat critically on this strong interpretation. The exhortation is to love everyone: family, friend, nodding acquaintance, and enemy, and apparently no distinctions are to be drawn. Indeed, one is positively to forgive enemies, virtually without limit. In at least two respects, the evolutionist sees the ethics emerging from his or her theory as breaking with this command.

First, there is the (intentional) lack of discrimination. Who is your neighbour? Everyone! The unknown child in Ethiopia has as much a moral claim on you as your own child. This is unacceptable to the evolutionist, and the reason becomes apparent as soon as we delve a little more deeply into the actual mechanisms which are thought to produce altruism (and subsequent 'altruism'). Two such mechanisms stand out. We have encountered one already, namely kin selection. Because it is biologically advantageous that our relatives reproduce, we have evolved features and behaviours inclining us to help such reproduction. The reward comes, not in material gains, but in vicarious reproduction. The other supposed mechanism (for which there is sound empirical support) operates between non-relatives, and is known as 'reciprocal altruism'. As the name implies, the supposition is that altruism evolves as the result of a kind of exchange mechanism, where we gain more from receiving the help of others than it costs us in giving help in return. Here, material return to us (or our relatives) is important.

I think it highly improbable that the kinds of altruism produced by these two mechanisms would not (in some degree) reflect the differences between the mechanisms. In particular, one would expect to find stronger kinds of obligation between relatives than between non-relatives – 'blood is thicker than water' – because the biological benefits are surely stronger, or at least more certain. A gene reproduced has a definite biological cash value. And as one encounters people further from one's immediate circle one would expect the sense of obligation to fall away. The possibilities of reciprocation begin to fade. (Wilson (1978) discusses this point in some detail.)

I must emphasize (to anticipate criticism) that I am now

talking of *moral* obligation. No one denies that, for obvious biological reasons, you love your own children more than those of others. My claim is that these feelings will be backed by different degrees of moral sentiment. Nor should you let all of the talk of reciprocation delude you into thinking morality is impossible. I believe that the reciprocation is enforced by morality! A person can ask help of another, not in return for help offered, but because it is right that the other extend help. You have an obligation to help me, and if you do not, then (in the name of morality) I and others regard you as beyond the pale, in some sense.

Furthermore, do note that I am not saying, callously, that we have no obligations at all to the Third World. Of course we have. It is just that most people feel that charity begins at home: the obligations of Americans are greater to the children of the ghetto than to the children of the desert, because the former are in the American community to a degree that the latter are not.

The evolutionist breaks from the stronger version of the love commandment in another way, also. Whilst there are undoubted biological virtues in not hastily pressing for instant retribution against every slight, real and imagined, the reciprocation demanded by biology surely rules out an unlimited willingness to turn the other cheek (Mackie 1978). Nor is it likely that the moral sense produced by evolution would demand this of you. Such obviously maladaptive behaviour could never have been produced and cherished by natural selection. There would have to be some early point at which abuse could be frustrated and barred. And one rather expects morality to back this frustration, rather than permitting ongoing personal attack.

My claim, therefore, is that if the Christian ethic is understood as being based on the stronger version of the love commandment, then there is conflict with the implications of modern evolutionary thought. It is not my intention here to decide between the opposing claims. Ultimately, that is for others to decide. However, apart from noting the above-mentioned doubts by scholars as to the status of the stronger interpretation, I would point out also that, independently, there are good reasons for opting for the evolutionist's position.

Most importantly, the minimum point for accepting a moral dictate has to be its inherent appeal or plausibility to the individual. 'Be kind to cabbages on Friday' would never be accepted because it seems so ridiculous. More than that, it is inconsistent with the attitudes that we have in other moral matters. At the most trivial level, for instance, morality is not the sort of thing that seems to come in and out of force on different days of the week. (What I am doing here is suggesting that one must, at a minimum, get one's moral beliefs into some reasonably harmonious set – what John Rawls (1971) describes as a state of 'reflective equilibrium'.)

In like fashion, I suggest that the evolutionist's understanding of morality accords much more with common intuitions and practices than does the strong interpretation. People do think that they have a special obligation to their families, and they likewise usually think that their first duties are to the poor of their own lands. People are far less likely to demand returns, even in the name of morality, of their children than of others. ('All you owe to me is that you do as much for your children as I have done for you.')

Furthermore, most people would think it quite irresponsible to let someone else sin against them 490 times. Long before this, the transgressor would have to be stopped. But, note how the very stopping itself would (as before) be done in the name of morality. You may well go on loving the sinner (even saying, you have an obligation to love the sinner); but you would claim that 'for their own good, as well as that of society, they ought to be constrained.' To think otherwise would be an abrogation of your moral duties.

At this point, perhaps, the response by the would-be strong interpretor is one invoking a rather extreme version of the Divine Command theory of ethics. It will be argued that our every-day moral intuitions are relatively limited. But, thanks to the words of Jesus, we now know that God demands of us very much more. Hence, our obligation as practising, believing Christians is to follow His will – blindly, if necessary. We have forced upon us what Kierkegaard (in discussing Abraham's preparedness to sacrifice Isaac) called 'a teleological suspension of the ethical' (Green 1982).

In company with many Christians, I feel uncomfortable with a god who demands of us (what our nature leads us to regard as) the morally perverse. However, this response does point to the major area of ethical enquiry which yet remains to be discussed, namely that pertaining to foundations. Accordingly, to end this enquiry, I turn now to this topic.

ULTIMATE FOUNDATIONS

I have been looking at the kinds of ethical claims made on the bases of evolution and Christianity, respectively. (This is known as 'substantive' or 'normative' ethics.) It is necessary now to look at the grounds on which such claims are made: so-called 'meta-ethics'. I suspect that many would argue that, if the evolutionist attempts to say anything interesting or significant, he or she runs into immediate problems. Allowing that the ethical sense has evolved in the way sketched above, supposedly (that is, in the eyes of the critic) this says nothing of reasoned justification. Grant that it is true (in either weaker or stronger sense) that we should love our neighbours as ourselves. The supposition that this is supported by the evolutionary process does indeed commit the naturalistic fallacy with a vengeance (Singer 1981).

I am sensitive to comments such as these; but I maintain that they miss the full import of evolutionism. Darwinian theory does speak to foundations, albeit in a negative sense. My claim is that the recognition of morality as merely a biological adaptation shows that there can be no foundation of the kind traditionally sought, whether by evolutionists, Christians, or others! I do not mean that ethics is a total chimera, for it obviously exists in some sense. But I do claim that, considered as a rationally justifiable set of claims about an objective something, it is illusory (Ruse 1986b). I appreciate that when somebody says 'Love thy neighbour as thyself', they think they are referring above and beyond themselves. It is a binding commandment, unlike a mere subjective claim (like 'I like spinach and I hope you do too'). Nevertheless, to a Darwinian evolutionist it can be seen that such reference is truly without foundation. Morality is just an aid to survival and reproduction, and has no being beyond or without this.

Why should humans be thus deceived about the presumed objectivity of moral claims? The answer is easy to see. Unless we think morality is objectively true – a function of something outside of and higher than ourselves – it would not work. If I think I should help you when and only when I want to, I shall probably help you relatively infrequently. But, because I think I *ought* to help you – because I have no choice about my obligation, it being imposed upon me – I am much more likely, in fact, to help you. And, conversely (Mackie 1978, 1979). Hence, by its very nature, ethics is and has to be something which is, apparently, objective, even though we now know that, truly, it is not.

Clearly, here, the evolutionist and the Christian part company. Admittedly, there is no unanimity among Christians as to the true foundations of morality. While some subscribe to a divine command theory, others (no doubt impressed by arguments which go back to Plato's *Euthyphro*) would argue that there are independent standards of right and wrong to which even God subscribes. But, be this as it may, the Christian is surely committed to an independent, objective, moral code – a code which, ultimately, is unchanging, and not dependent on the contingencies of human nature. (Of course, like any moralist, the Christian appreciates that different times and different places call for different applications of this code.)

This independence is expressly denied by the Darwinian evolutionist. Morality is an ephemeral product of the evolutionary process, just as are other adaptations. It has no existence or being beyond this, and any deeper meaning is illusory (although put on us for good biological reasons). Yet is this not too quick a conclusion? Consider an analogy. We see the moving train with our sense organs, which clearly show their adaptive value as we step smartly out of the train's path. But no one would deny that the train genuinely exists, whether we see it or not. Perhaps, therefore, the same can and should be said of morality. It is true that our awareness of right and wrong depends on evolved organs, and that such awareness has adaptive value, but this is not to deny the independent existence of moral standards (Nozick 1981).

Unfortunately, however, the analogy breaks down. Consider two separate worlds, identical except that one has an objective morality and the other does not. Humans could have evolved in both worlds, to believe in exactly the same things! The two identical species could share thoughts about right and wrong. To suppose otherwise, that is, to suppose that only the world of objective morality could have humans believing in it, is to suppose an extra-scientific channelling of events – a channelling which is quite antithetical to modern evolutionism. In short, therefore, in a sense, the objective morality is redundant. Its existence is irrelevant to human thought and action. (Things are quite otherwise with the moving train. I can imagine two worlds, different in that one of them does not contain large, fast-moving, life-threatening objects. Human evolution might have been quite different in the two cases.)

The paradoxical nature of this conclusion hardly needs stressing. God wants you to be good; but His wishes and the existence of an independent morality are quite irrelevant to whether you will think you should be good, and whether you are good. In fact, the situation is even worse than this. Suppose we had evolved in a rather different way. Suppose, to take an extreme example, we had evolved from termite-like creatures, rather than from savannah-dwelling primates. Termites need to eat each other's faeces, in order to regain certain parasites used in digestion, which are lost during the termites' periodic moults. With such a background as this, our highest ethical imperatives might be very strange indeed. We would live our lives in blissful ignorance of what God or objective morality truly willed.

Nor can the Christian save the day by taking a Kantian approach, arguing that objectivity in ethics does not imply something 'out there', but is rather the condition which obtains when rational beings interact socially. Under this conception, ethics is regarded as a necessarily emerging relational product, rather as the truth of Pythagoras' theorem emerges as a property of right-angled triangles. Such a position might seem to save the day for the believer, for there is now no claim about independent standards, beyond and apart from actual, existing humans (Kant 1959, Rawls 1980).

Nevertheless, although I appreciate moves in this direction, I doubt the Kantian approach will do all that is required. Given social animals and the laws of nature, no doubt some form of reciprocation is demanded. Unfortunately, this reciprocation does not necessarily require morality as we know it. We have seen that 'altruism' does not presuppose altruism. Suppose, for instance, we all thought like the so-called Superpowers. One side dislikes and distrust the other, feeling that is has an *obligation* to behave this way and to like only its own side. A way of existing together is thereby achieved, in itself no less rational than our present state. You might even say that it is more rational. Yet nothing resembling Christian morality is to be found. John Foster Dulles might have been very efficient at dealing with the Russians. He did not love them as himself.

Clearly, something has gone badly wrong, particularly when you reflect that, already, our evolution might have taken us away from the supposed true nature of morality. Perhaps we really ought to hate our neighbours, but we, poor fools, think otherwise! This consequence is absurd – to evolutionist and Christian alike.

CONCLUSION

Right at the end, therefore, I am forced to conclude that the following-through of the implications of modern evolutionary theory causes severe problems for the practising Christian. I agree that most of the supposed road-blocks to faith are less than troublesome. Indeed, in many respects I believe that a full understanding of modern evolutionary theory even helps to solve problems which still worry many Christians considerably. I refer, particularly, to the realm of sexual behaviour. Furthermore, even with the Christian's central love commandment, I suggest that there is much in evolutionism acceptable to the majority of believers, although, of course, I would not pretend that everyone would be happy with my conclusions. (But, what position would be willingly embraced by all Christians?)

However, when it comes to ultimate foundations, the evolutionist and the Christian part company. For the evolutionist, morality – that which yields standards of right and wrong – rests in the contingencies of human nature. In an important sense, therefore there are no ultimate foundations, just a biological illusion of objectivity. For the Christian, morality simply has to be something more than this, even though the precise nature of 'more' would be unpacked differently by different people.

This gap between evolutionist and Christian should not be minimized. If the Christian points out that it just so happens that what he or she believes by relevation ('Love thy neighbour') coincides with what the evolutionist finds emerging from the natural process ('Love thy neighbour'), the evolutionist protests that this is altogether too much of coincidence to be shrugged off. That we humans should just so have happened to have evolved to that very morality which is endorsed by God, imputes a teleological flavour to the course of evolution which is alien to modern science.

I do not say that no reconciliation could or should be sought. One might, for instance, argue that God so arranged natural laws and the initial ordering of matter, that the correct morality was bound to emerge. However, this kind of predetermination is certainly not acceptable to all Christians. To say the least, such a view puts in shadow the effectiveness of human freedom, not to mention its making God directly responsible for the ongoing cruelties which accompany the struggle for existence.

But I say nothing now that has not been discussed by Christians, for generations. I simply conclude by reiterating the tensions I see between the evolutionist's understanding of morality and the claims of the Christian. That troublesome relationship between science and faith seems still to be with us.

NOTE

1. The point is not that there is a contradiction between evolution and Christianity, but that – evolution giving us no moral rules – Christianity calls upon us to rise above our brute natures.

BIBLIOGRAPHY

Alexander, R.D. (1971) 'The search for an evolutionary philosophy', *Proceedings of the Royal Society of Victoria, Australia*, 84: 99-120.
— (1974) 'The evolution of social behavior', *Annual Review of Ecology and Systematics* 5: 325-84.
— (1975) 'The search for a general theory of behavior', *Behavioral Science* 20: 77-100.
— (1977a) 'Evolution, human behaviour, and determinism ', in F. Suppe and P. Asquith (eds) *PSA 1976*, East Lansing: Philosophy of Science Association.
— (1977b) 'Natural selection and the analysis of human sociality', in C.E. Goulden (ed.) *Changing Scenes in Natural Sciences*, Philadelphia: Philadelphia Academy of Natural Sciences.
— (1979) *Darwinism and Human Affairs*, Seattle: University of Washington Press.
Allen, E. *et al* (1975) letter to editor, *New York Review of Books* 22, 18: 43-4.
— (1976) 'Sociobiology: a new biological determinism', *BioScience* 26: 182-6.
— (1977) 'Sociobiology: a new biological determinism', in Sociobiology Study Group of Boston (ed.) *Biology as a Social Weapon*, Minneapolis: Burgess.
Appel, T.A. (1987) *The Cuvier-Geofroy Debate: French Biology in the Decades before Darwin*, New York: Oxford University Press.
Aquinas, Saint Thomas (1968) *Summa Theologiae, 43, Temperance* (2a, 2ae: 151-4), trans. T. Gilby, London: Blackfriars.
Arnold, A.J. and Fristrup, K. (1982) 'The theory of evolution by natural selection: a hierarchical expansion', *Paleobiology* 8:111-29. Reprinted in R.N. Brandon and R.M. Burian (eds) *Genes, Organisms and Populations*, Cambridge, Mass: MIT Press: 292-320.
Atwater, T. (1973) letter to A. Cox, in A. Cox (ed.) *Plate Tectonics and Geomagnetic Reversals*, San Francisco: Freeman: 535-6.
Ayala, F. (1974) Introduction to F. Ayala and Th. Dobzhansky (eds) *Studies in the Philosophy of Biology*, Berkeley: University of California

Press.
— (1985) 'The theory of evolution: recent successes and challenges', in E. McMullin (ed.) *Evolution and Creation*, Notre Dame: University of Notre Dame Press: 59-90.
— and Dobzhansky, Th. (eds) (1974) *Studies in the Philosophy of Biology*, Berkeley: University of California Press.
— and Valentine, J.W. (1979) *Evolving*, Menlo Park, California: Benjamin/Cummings.
Ayers, M.R. (1981) 'Locke vs. Aristotle on natural kinds', *Journal of Philosophy* 78: 247-72.
Barash, D.P. (1977) *Sociobiology and Behavior*, New York: Elsevier.
Barlow, N. (1967) *Darwin and Henslow: The Growth of an Idea. Letters, 1831-1860*, Berkeley: University of California Press.
Barrett, P.H., Gautrey, P.J., Herbert, S., Kohn, D., and Smith, S. (eds) (1987) *Charles Darwin's Notebooks, 1836-1844*, Ithaca, NY: Cornell University Press.
Bateson, P. (1976) 'Specificity and the origins of behavior', *Advances in the Study of Behavior* 6: 1-20.
— (1983) 'Rules for changing the rules', in D.S. Bendall (ed.) *Evolution from Molecules to Men*, Cambridge: Cambridge University Press.
— (1986a) 'Sociobiology and human politics', in S. Rose and L. Appignanesi (eds) *Science and Beyond*, Oxford: Blackwell.
— (1986b) 'A biological perspective on human sociobiology', MS for Liberty Fund conference on evolution and ethics.
Beatty, J. (1980) 'Optimal-design models and the strategy of model building in evolutionary biology', *Philosophy of Science* 47: 532-61.
— (1982) 'Classes and cladists', *Systematic Zoology* 31: 25-34.
Beck, L.W. (1972) 'Extraterrestrial intelligent life', *Proceedings and Addresses of the American Philosophical Association* XLV: 5-21.
Beckner, M. (1959) *The Biological Way of Thought*, New York: Columbia University Press.
— (1969) 'Function and teleology', *Journal of the History of Biology* 2: 151-64.
Berkeley, G. (1963) *Works on Vision*, Indianapolis: Library of Liberal Arts.
Betz, D. (1985) *Essays on the Sermon on the Mount*, Philadelphia: Fortress.
Betzig, L.L., Borgerhoff Mulder, M., and Turke, P.W. (eds) (1988) *Human Reproductive Behavior*, Cambridge: Cambridge University Press.
Billingham, J. (1981) preface to J. Billingham (ed.) *Life in the Universe*, Cambridge, Mass: MIT Press.
Black, M. (1962) *Models and Metaphors*, Ithaca, NY: Cornell University Press.
Bogess, J. (1979) 'Troop male membership changes and infant killing in langurs', (*Presbytis entellus*), *Folia Primatologica* 32: 65-107.
Boorman, S.A. and Levitt, P.R. (1972) 'Group selection on the boundary of a stable population', *Proceedings of the National*

Academy of Sciences, USA 69(9): 2711-13.
—— (1973) 'Group selection on the boundary of a stable population', *Theoretical Population Biology* 4 (1): 82-128.
Borgerhoff Mulder, M. (1987a) 'Reproductive success of three Kipsigis cohorts', in T.H. Clutton-Brock (ed.) *Reproductive Success: Studies of Selection and Adaptation in Contrasting Breeding Systems*, Chicago: University of Chicago Press.
—— (1987b) 'Adaptation and evolutionary approaches to anthropology', *Man* 22: 25-41.
—— (1987c) 'On cultural and reproductive success', Kipsigis trial evidence, MS.
—— (1988) 'Kipsigis bridewealth payments', in L.L. Betzig, M. Borgerhoff Mulder, and P.W. Turke (eds) *Human Reproductive Behavior: A Darwinian Perspective*, Cambridge: Cambridge University Press: 65-82.
Bowler, P. (1976) *Fossils and Progress*, New York: Science History Publications.
Boswell, J. (1980) *Christianity, Social Tolerance, and Homosexuality*, Chicago: Chicago University Press.
Boyd, R. and Richerson, P. (1985) *Culture and the Evolutionary Process*, Chicago: University of Chicago Press.
Brandon, R.N. and Burian, R.M. (eds) (1984) *Genes, Organisms, Populations: Controversies Over the Units of Selection*, Cambridge, Mass: MIT Press.
Brewster, D. (1838) 'Review of Comte, *Cours de Philosophie Positive*', *Edinburgh Review* 67: 271-308.
—— (1854) *More Worlds than One: The Creed of the Philosopher and the Hope of the Christian*, London: Murray.
Brooke, J.H. (1977) 'Natural theology and the plurality of worlds: Observations on the Brewster-Whewell debate', *Annals of Science* 34: 221-86.
Brownmiller, S. (1975) *Against Our Will: Men, Women, and Rape*, New York: Simon & Schuster.
Buchdahl, G. (1971) 'Inductivist *versus* deductivist approaches in the philosophy of science as illustrated by some controversies between Whewell and Mill', *Monist* 55: 343-67.
Buckland, W. (1836) *Geology and Mineralogy: Bridgewater Treatise 6*, London: Pickering.
Bullard, E., Everett, J. and Smith, A.G. (1965) 'The fit of the continents around the Atlantic', in P.M.S. Blackett, E. Bullard and S.K. Runcorn (eds) 'A symposium on continental drift', *Philosophical Transactions of the Royal Society of London* A, 258, 1088: 41-51.
Bultmann, R. (1951) *Theology of the New Testament*, New York: Charles Scribner's Sons.
Burchfield, J.D. (1974) 'Darwin and the dilemma of geological time', *Isis* 65: 300-21.
Butts, R.E. (1968) *William Whewell's Theory of Scientific Method*,

Pittsburgh: University of Pittsburgh Press.
— (1970) 'Whewell on Newton's rules of philosophizing', in R.E. Butts and J. Davis (eds) *The Methodological Heritage of Newton*, Toronto: University of Toronto Press.
Calvin, J. (1962) *Institutes of the Christian Religion*, Grand Rapids: Eerdmans.
Cannon, W.F. (1961) 'John F.W. Herschel and the idea of science', *Journal of the History of Ideas* 22: 215-39.
Caplan, A. (1980) 'Have species become declassé?', in P. Asquith and R. Giere (eds) *PSA 1980*, East Lansing, Mich: Philosophy of Science Association: 1: 71-82.
— (1981) 'Back to class: a note on the ontology of species', *Philosophy of Science* 48: 130-40.
Caro, T.M. and Borgerhoff Mulder, M. (1987) 'The problem of adaptation in the study of human behaviour', *Ethology and Sociobiology* 8: 61-72.
Cavalli-Sforza, L.L. and Feldman, M.W. (1981) *Cultural Transmission and Evolution*, Princeton: Princeton University Press.
Chamberlin, R.T. (1928) 'Some of the objections to Wegener's theory', in W.A.J.M. van Waterschoot van der Gracht, *et al.* (eds) *Theory of Continental Drift*, Tulsa: American Association of Petroleum Geologists: 83-7.
Chambers, R. (1844) *Vestiges of the Natural History of Creation*, London: Churchill.
Chomsky, N. (1957) *Syntactic Structures*, The Hague: Mouton.
Clutton-Brock, T.H., Guiness, F.E. and Albon, S.D. (1982) *Red Deer: Behavior and Ecology of the Two Sexes*, Chicago: University of Chicago Press.
Conan Doyle, A. (1980) *The Complete Sherlock Holmes*, Harmondsworth: Penguin Books.
Cox, A. (ed.) (1973) *Plate Tectonics and Geomagnetic Reversals*, San Francisco: Freeman.
Crook, J.M. (1980) *The Evolution of Human Consciousness*, Oxford: Oxford University Press.
Darwin, C. (1842) *The Structure and Distribution of Coral Reefs*, London: Smith Elder.
— (1851) *A Monograph of the Sub-class Cirripedia, with Figures of all the Species. The Lepadidae; or Pedunculated Cirripedes*, London: Ray Society.
— (1854) *A Monograph of the Sub-class Cirripedia, with Figures of all the Species. The Balanidae (or Sessile Cirripedes); the Verrucidae, &c.*, London: Ray Society.
— (1859) *On the Origin of Species*, London: Murray.
— (1862) 'On the two forms, or dimorphic condition, in the species of *Primula*, and on their remarkable sexual relations', *Journal of the Proceedings of the Linnean Society (Botany)* 6: 77-96. Reprinted in P.H. Barret (ed.) *The Collected Papers of Charles Darwin*, Chicago: University of Chicago Press: 2, 45-63.

— (1865) On the sexual relations of the three forms of *Lythrum salicavia*, *Journal of the Proceedings of the Linnean Society (Botany)* 8: 169-96. Reprinted in P.H. Barrett (ed.) *The Collected Papers of Charles Darwin*, Chicago: University of Chicago Press: 2, 106-31.

— (1868) *The Variation of Animals and Plants under Domestication*, London: Murray.

— (1871) *Descent of Man*, London: Murray.

— (1959) *On the Origin of Species*, Variorum Text, M. Peckham (ed.). Philadelphia: University of Pennsylvania Press.

— (1969) *Autobiography*, N. Barlow (ed.), New York: Norton.

— (1975) *Natural Selection*, R.C. Stauffer (ed.), Cambridge: Cambridge University Press.

— and Wallace, A.R. (1958) *Evolution by Natural Selection*, G. de Beer (ed.), Cambridge: Cambridge University Press.

Darwin, F. (ed.) (1887) *The Life and Letters of Charles Darwin, Including an Autobiographical Chapter*, London: Murray.

— and Seward, A.C. (ed.) (1903) *More Letters of Charles Darwin*, London: Murray.

Dawkins, R. (1976) *The Selfish Gene*, Oxford: Oxford University Press.

— (1981) 'In defence of selfish genes', *Philosophy* 56: 556-73.

— (1982) 'Replicators and vehicles', in King's College Sociobiology Group (eds) *Current Problems in Sociobiology*, Cambridge: Cambridge University Press: 45-64.

— (1983) 'Universal Darwinism', in D.S. Bendall (ed.) *Evolution from Molecules to Men*, Cambridge: Cambridge University Press: 403-25.

— (1986) *The Blind Watchmaker*, London: Longman.

de Beer, G. (ed.) (1960-7) 'Darwin's notebooks on transmutation of species', *Bulletin of the British Museum (Natural History) Historical Series* 2: 27-200; 3: 129-76.

de Waal, F. (1982) *Chimpanzee Politics*, London: Jonathan Cape.

Dick, S. (1982) *Plurality of Worlds: The Origins of the Extraterrestrial Life Debate from Democritus to Kant*, Cambridge: Cambridge University Press.

Dickerson, R.E. (1978) 'Chemical evolution and the origin of life', *Scientific American*, September: 70-86.

Dobzhansky, Th. (1951) *Genetics and the Origin of Species*, 3rd ed., New York: Columbia University Press.

— (1962) *Mankind Evolving*, New York: Columbia University Press.

— (1970) *Genetics of the Evolutionary Process*, New York: Columbia University Press.

— Ayala, F.J., Stebbins, G.L., and Valentine, J.W. (1977) *Evolution*, San Francisco: Freeman.

Ducasse, C.J. (1960a) 'John F.W. Herschel's methods of experimental inquiry', in R.M. Blake (ed.) *Theories of Scientific Method*, Seattle: University of Washington Press.

— (1960b) 'William Whewell's Philosophy of Scientific Discovery', in R.M. Blake (ed.) *Theories of Scientific method*, Seattle: University of Washington Press.

Dupré, J. (1981) 'Natural kinds and biological taxa', *Philosophical Review* 90: 66-90.

Durant, J. (ed.) (1985) *Darwinism and Divinity*, Oxford: Blackwell.

Durham, W.H. (1976) 'The adaptive significance of cultural behavior', *Human Ecology* 4: 89-121.

du Toit, A.L. (1937) *Our Wandering Continents*, Edinburgh: Oliver & Boyd.

Eigen, M., Gardiner, W., Schuster, P., and Winkler-Oswatitsch, R. (1981) 'The origin of genetic information', *Scientific American* 244 (4): 88-118.

Ellegard, A. (1958) *Darwin and the General Reader*, Göteborg: Göteborgs Universitets Arsskrift.

Eldredge, N. (1971) 'The allopatric model and phylogeny in Paleozoic invertebrates', *Evolution* 25: 156-67.

— (1984) 'Large scale biological entities and the evolutionary process', in P. Asquith and P. Kitcher (eds) *PSA 1984*, East Lansing, Mich: Philosophy of Science Association: 2, 525-42.

— (1985a) *Time Frames: The Rethinking of Darwinian Evolution and the Theory of Punctuated Equilibria*, New York: Simon & Schuster.

— (1985b) *Unfinished Synthesis. Biological Hierarchies and Modern Evolutionary Thought*, Oxford: Oxford University Press.

— and Gould, S.J. (1972) 'Punctuated equilibria: an alternative to phyletic gradualism', in T.J.M. Schopf (ed.) *Models in Paleobiology*. San Francisco: Freeman, Cooper.

— and Cracraft, J. (1980) *Phylogenetic Patterns and the Evolutionary Process*, New York: Columbia University Press.

Endler, J.A. (1977) *Geographic Variation, Speciation, and Clines*, Princeton: Princeton University Press.

Engels, F. (1971) *Dialectics of Nature*, Moscow: Progress Publishers.

Farley, J. (1977) *The Spontaneous Generation Controversy: From Descartes to Oparin*, Baltimore: Johns Hopkins Press.

Feduccia, A. (1980) *The Age of Birds*, Cambridge, Mass: Harvard University Press.

Fisher, R.A. (1930) *The Genetical Theory of Natural Selection*, revised and reprinted 1958, New York: Dover.

Flew, A.G.N. (1959) 'The structure of Darwinism', *New Biology* 28: 18-34.

— (1967) *Evolutionary Ethics*, London: Macmillan.

Forbes, J. and King, J. (eds) (1982) *Primate Behavior*, New York: Academic Press.

Frankel, H. (1979) 'The career of continental drift theory: an application of Imré Lakatos' analysis of scientific growth to the rise of drift theory', *Studies in History and Philosophy of Science* 10: 21-66.

Freud, S. (1913) *Totem and Taboo*, in J. Strachey (ed.) *Collected Works of Freud*, 13, (1953) London: Hogarth.

Futuyma, D. (1979) *Evolutionary Biology*, Sunderland, Mass: Sinauer.

Ghiselin, M. (1966) 'On psychologism in the logic of taxonomic controversies', *Systematic Zoology* 15: 207-15.

— (1969) *The Triumph of the Darwinian Method*, Berkeley: University of California Press.

— (1974a) *The Economy of Nature and the Evolution of Sex*, Berkeley: University of California Press.

— (1974b) 'A radical solution to the species problem', *Systematic Zoology* 23: 536-44.

— (1981) 'Categories, life, and thinking', *Behavior and Brain Sciences* 4: 269-313.

— (1987) 'Species concepts, individuality, and objectivity', *Biology and Philosophy* 2: 127-45.

Gilluly, J., Waters, A., and Woodford, A. (1959) *Principles of Geology*, San Francisco: Freeman, 4th ed. (1974).

— (1971) 'Plate tectonics and magmatic evolution', *Bulletin of the Geological Society of America*. 82: 2382-96, reprinted in A. Cox (ed.) (1973) *Plate Tectonics and Geomagnetic Reversals*, San Francisco: Freeman: 648-58.

Gingerich, P.D. (1976) 'Paleontology and phylogeny: patterns of evolution at the species level in early Tertiary mammals', *American Journal of Science* 276: 1-28.

— (1977) 'Patterns of evolution in the mammalian fossil record', in A. Hallam (ed.) *Patterns of Evolution, as Illustrated by the Fossil Record*, Amsterdam: Elsevier: 469-500.

Ginsberg, M. (1953) *Evolution and Progress*, London: Heinemann.

Gish, D.T. (1973) *Evolution: The Fossils Say No!* San Diego: Creation-Life.

Goldschmidt, R.B. (1940) *The Material Basis of Evolution*, New Haven: Yale University Press.

Gould, S.J. (1971) 'D'Arcy Thompson and the science of form', *New Literary History* 2: 229-58.

— (1973) 'Positive allometry of antlers in the "Irish Elk", *Megaloceros giganteus*', *Nature* 244: 375-6.

— (1977a) *Ontogeny and Phylogeny*, Cambridge, Mass: Harvard University Press.

— (1977b) *Ever Since Darwin*, New York: Norton.

— (1978) 'Sociobiology: the art of story-telling', *New Scientist* 80: 530-3.

— (1979a) 'A quahog is a quahog', *Natural History* 88: 18-26.

— (1979b) 'Episodic change versus gradualist dogma', *Science and Nature* 2: 5-12.

— (1979c) 'Agassiz's marginalia in Lyell's *Principles*, or the perils of uniformity and the ambiguity of heroes', *Studies in the History of Biology* 3: 119-38.

— (1980a) 'The promise of paleobiology as a nomothetic, evolutionary discipline', *Paleobiology* 6: 96-118.

— (1980b) 'Is a new and general theory of evolution emerging?', *Paleobiology* 6: 119-30.

— (1980c) *The Panda's Thumb*, New York: Norton.

— (1981) *The Mismeasurement of Man*, New York: Norton.

279

BIBLIOGRAPHY

— (1982a) 'Punctuated equilibrium – a different way of seeing', in J. Cherfas (ed.) *Darwin Up to Date*, London: IPC Magazines: 26-30.
— (1982b) 'The meaning of punctuated equilibrium and its role in validating a hierarchical approach to macroevolution', in R. Milkman (ed.) *Perspectives On Evolution*, Sunderland, Mass: Sinauer Associates: 83-104.
— (1982c) 'Darwinism and the expansion of evolutionary theory', *Science* 216: 380-7.
— (1983a) 'Irrelevance, submission, and partnership: the changing role of palaeontology in Darwin's three centennials, and a modest proposal for macroevolution', in D.S. Bendall (ed.) *Evolution from Molecules to Men*, Cambridge: Cambridge University Press: 347-66.
— (1983b) 'The hardening of the modern synthesis', in M. Grene (ed.) *Dimensions of Darwinism*, Cambridge: Cambridge University Press: 71-93.
— (1983c) *Hen's Teeth and Horses' Toes*, New York: Norton.
— and Eldredge, N. (1977) 'Punctuated equilibria: the tempo and mode of evolution reconsidered', *Paleobiology* 3: 115-51.
— and Eldredge, N. (1986) 'Punctuated equilibrium at the third stage', *Systematic Zoology* 35 (1): 143-8.
— and Lewontin, R. (1979) 'The spandrels of San Marco and the Panglossian paradigm', *Proceedings of the Royal Society of London* B205: 581-98.
Grant, V. (1971) *Plant Speciation*, New York: Columbia University Press.
— (1981a) *Plant Speciation*, 2nd ed., New York: Columbia University Press.
— (1981b) 'The genetic goal of speciation', *Biologisches Zentralblatt* 100: 473-82.
Green, R.M. (1982) 'Abraham, Isaac, and the Jewish tradition: an ethical reappraisal', *The Journal of Religious Ethics* 10: 1-21.
Griffin, D.A. (1976) *The Question of Animal Awareness: Evolutionary Continuity of Mental Experience*, New York: Rockefeller University Press.
Grossman, N. (1974) 'Empiricism and the possibility of encountering intelligent beings with different sense-structures', *Journal of Philosophy* 71: 815-21.
Gruber, H.E. and Barrett, P.H. (1974) *Darwin on Man*, New York: Dutton.
Hallam, A. (1973) *A Revolution in the Earth Sciences: From Continental Drift to Plate Tectonics*, Oxford: Clarendon Press.
— (ed.) (1977) *Patterns of Evolution as Illustrated by the Fossil Record*, Amsterdam: Elsevier.
Hamilton, W.D. (1964a) 'The genetical evolution of social behaviour, I', *Journal of Theoretical Biology* 7: 1-16.
— (1964b) 'The genetical evolution of social behaviour, II', *Journal of Theoretical Biology* 7: 17-32.
Haraway, D.J. (1983) 'Tne contest for primate nature: daughters of

280

man-the-hunter in the field, 1960-1980', in M.E. Kann (ed.) *The Future of American Democracy: Views from the Left*, Philadelphia: Temple University Press: 175-207.

Hart, M.H. and Zuckerman, B. (eds) (1982) *Extraterrestrials: Where Are They?*, New York: Pergamon.

Hartung, J. (1985) 'Matrilineal inheritance: new theory and analysis', *Behavioral and Brain Sciences* 8: 661-88.

Harvey, P.H. and Ralls, K. (1986) 'Do animals avoid incest?', *Nature* 320: 575-6.

Hempel, C.G. (1952) *Fundamentals of Concept Formation in Empirical Science*, Chicago: University of Chicago Press.

Henslow, J. (1835) *Descriptive and Physiological Botany*, London: Longman.

Herbert, S. (1971) 'Darwin, Malthus, and selection', *Journal of the History of Biology* 4: 209-17.

Herschel, J.F.W. (1831) *Preliminary Discourse on the Study of Natural Philosophy*, London: Longman, Rees, Orme, Brown, and Green.

— (1833) *Treatise on Astronomy*, London: Longman.

— (1841) 'History ... and Philosophy of the Inductive Sciences ... by William Whewell ...', *Quarterly Review* 135: 177-238.

— (1861) *Physical Geography*, Edinburgh: Black.

Hess, M.H. (1962) 'History of ocean basins', in A. Engel, H. James and B. Leonard(eds) *Petrologic Studies: A Volume to Honor A.F. Buddington*, New York: Geological Society of America: 599-620, reprinted in A. Cox (ed.) (1973) *Plate Tectonics and Geomagnetic Reversals*, San Francisco: Freeman: 23-38.

Hesse, M. (1966) *Models and Analogies in Science*, Notre Dame, Indiana: University of Notre Dame Press.

Hilton, B., Callahan, D., Harris, M., Condliffe, P., and Berkley, B. (eds) (1973) *Ethical Issues in Human Genetics*, New York: Plenum.

Himmelfarb, G. (1962) *Darwin and the Darwinian Revolution*, New York: Anchor.

Holmes, A. (1931) 'Radioactivity and earth movements', *Transactions of the Geological Society of Glasgow* 18: 559-606.

Holsinger, K.E. (1984) 'The nature of biological species', *Philosophy of Science* 51: 293-307.

Hopkins, W. (1860) 'Physical theories of the phenomenon of life', *Fraser's Magazine* 61: 739-52; 62: 74-90.

Hrdy, S.B. (1977) *The Langurs of Abu: Female and Male Strategies of Reproduction*, Cambridge, Mass: Harvard University Press.

— (1981) *The Woman that Never Evolved*, Cambridge, Mass: Harvard University Press.

Hubbard, R. (1983) 'Have only men evolved?', in S. Harding and M.B. Hintikka (eds) *Discovering Reality*, Dordrecht: Reidel, 45-69.

Hull, D.L. (1965) 'The effect of essentialism on taxonomy: two thousand years of stasis', *British Journal for the Philosophy of Science* 15: 314-26; 16: 1-18.

— (1972) 'Reduction in genetics—biology or philosophy?', *Philosophy*

of Science 39: 491-99.
— (ed.) (1973a) *Darwin and His Critics*, Cambridge, Mass: Harvard University Press.
— (1973b) 'Reduction in genetics–doing the impossible', in P. Suppes *et al.* (eds) *Logic, Methodology and Philosophy of Science* IV: 619-35.
— (1974) *Philosophy of Biological Science*, Englewood Cliffs: Prentice-Hall.
— (1975) 'Central subjects and historical narratives', *History and Theory* 14: 253-74.
— (1976a) 'Are species really individuals?', *Systematic Zoology* 25: 174-91.
— (1976b) 'Informal aspects of theory reduction', in R.S. Cohen and A. Michalos (eds) *PSA 1974*, Dordrecht: Reidel: 653-70.
— (1978a) 'Sociobiology: scientific bandwagon or traveling medicine show?', in M.S. Gregory, A. Silvers, and D. Sutch (eds) *Sociobiology and Human Nature*, San Francisco: Jossey-Bass: 136-63.
— (1978b) 'A matter of individuality', *Philosophy of Science* 45: 335-60.
— (1979) 'The limits of cladism', *Systematic Zoology* 28: 414-38.
— (1980) 'Sociobiology: another new synthesis', in W. Barlow and J. Silverberg (eds) *Sociobiology: Beyond Nature-Nurture?* Boulder, Co: Westview Press: 77-96.
— (1981) 'Kitts and Kitts and Caplan on species', *Philosophy of Science* 48: 141-52.
Hume, D. (1779) *Dialogues Concerning Natural Religion*, reprinted in R. Wollheim (ed.) (1963) *Hume on Religion*, London: Collins.
Huxley, J. (1942) *Evolution: The Modern Synthesis*, London: Allen & Unwin.
— (1947) *Evolution and Ethics*, London: Pilot.
Huxley, T.H. (1860) 'The origin of species', reprinted in T.H. Huxley (1893) *Darwiniana*, London: Macmillan.
— (1894) *Evolution and Ethics*, London: Macmillan.
Irons, W.G. (1979) 'Cultural and biological success', in N. Chagnon and W. Irons (eds) *Evolutionary Biology and Human Social Behavior: An Anthropological Perspective*, North Scituate, Mass: Duxbury: 257-72.
Isaac, G. (1983) 'Aspects of human evolution', in D.S. Bendall (ed.) *Evolution from Molecules to Men*, Cambridge: Cambridge University Press: 509-43.
Isacks, B., Oliver, J., and Sykes, L. (1968) 'Seismology and the new global tectonics', *Journal of Geophysical Resources* 73: 5855-99, reprinted in A. Cox (ed.) (1973) *Plate Tectonics and Geomagnetic Reversals*, San Francisco: Freeman: 358-400.
Johanson, D. and Edey, M. (1981) *Lucy: The Beginnings of Humankind*, New York: Simon & Schuster.
— and White, T.D. (1979) 'A systematic assessment of early African hominids', *Science* 203: 321-30.
Judson, M.F. (1979) *The Eighth Day of Creation*, New York: Simon & Schuster.

Kant, I. (1959) *Foundations of the Metaphysics of Morals*, trans. L.W. Beck, Indianapolis: Bobbs-Merrill.

— (1980) *Critique of Judgement*, trans. J. Meredith, Oxford: Clarendon Press.

Kimura, M. (1983) *The Neutral Theory of Molecular Evolution*, Cambridge: Cambridge University Press.

Kitcher, P. (1984) 'Species', *Philosophy of Science* 51: 308-35.

— (1985) *Vaulting Ambition*, Cambridge, Mass: MIT Press.

Kitts, D.B. (1974) 'Continental drift and scientific revolution', *American Association of Petroleum Geologists Bulletin*, reprinted in D.B. Kitts (1977) *The Structure of Geology*, Dallas: Southern Methodist University Press: 115-27.

— and Kitts, D.J. (1979) 'Biological species as natural kinds', *Philosophy of Science* 46: 613-22.

Koertge, N. (1984) *Valley of the Amazons*, New York: St Martin's Press.

Kohlberg, L. (1969) 'Stages and sequence: the cognitive-developmental approach to socialization', in I.A. Goslin (ed.) *Handbook of Socialization Theory and Research*, Chicago: Rand McNally: 347-80.

Körner, S. (1960) 'On philosophical arguments in physics', in E.H. Madden (ed.) *The Structure of Scientific Thought*, Boston: Houghton Mifflin: 106-10.

Kottler, M.J. (1974) 'Alfred Russel Wallace, the origin of man, and spiritualism', *Isis* 65: 145-92.

— (1976) 'Isolation and Speciation, 1837-1900', unpublished Yale Ph.D. thesis.

Kripke, S.A. (1972) 'Naming and necessity', in D. Davidson and G. Harman (eds) *Semantics of Natural Language*, Dordrecht: Reidel: 253-355.

Kuhn, T.S. (1957) *The Copernican Revolution*, Cambridge: Mass: Harvard University Press.

— (1962) *The Structure of Scientific Revolutions*, Chicago: University of Chicago Press (2nd ed. 1970).

Lack, D. (1954) *The Natural Regulation of Animal Numbers*, Oxford: Oxford University Press.

— (1966) *Population Studies of Birds*, Oxford: Oxford University Press.

Lande, R. (1980) 'Genetic variation and phenotypic evolution during allopatric speciation', *American Naturalist* 116: 463-79.

Laudan, L. (1971) 'William Whewell on the consilience of inductions', *The Monist* 55: 368-91.

— (1977) *Progress and Its Problems*, Berkeley: University of California Press.

Laudan, R. (1976) 'William Smith. Stratigraphy without palaeontology', *Centaurus* 20: 210-26.

Leplin, J. (ed.) (1984) *Scientific Realism*, Berkeley: University of California Press.

Levin, D.A. (1979) 'The nature of plant species', *Science* 204: 381-4.

— (1981) 'Dispersal versus gene flow in plants', *Annals of the Missouri Botanical Garden* 68: 233-53.

Levins, R. (1970) 'Extinction', in M. Gerstenhaber (ed.) *Some Mathematical Questions in Biology*, Providence: American Mathematical Society: 77-107.

Lévi-Strauss, C. (1969) *The Elementary Structures of Kinship*, Boston: Beacon Press.

Lewin, R. (1980) 'Evolutionary theory under fire', *Science* 210: 883-7.

Lewontin, R.C. (1970) 'The units of selection', *Annual Review of Ecology and Systematics* 1: 1-18.

— (1974) *The Genetic Basis of Evolutionary Change*, New York: Columbia University Press.

— (1978) 'Adaptation', *Scientific American* 235 (3): 212-30.

— and Levins, R. (1976) 'The problem of Lysenkoism', in H. and S. Rose (eds) *The Radicalisation of Science*, London: Macmillan: 32-64.

Limoges, C. (1970) *La Sélection naturelle*, Paris: Universitaires de France.

Locke, J. (1975) *An Essay Concerning Human Understanding*, P.H. Nidditch (ed.), Oxford: Oxford University Press.

Lorenz, K. (1966) *On Aggression*, London: Methuen.

Lovejoy, C.O. (1981) 'The origin of man', *Science* 211: 341-50.

Lumsden, C. and Wilson, E.O. (1981) *Genes, Mind, and Culture*, Cambridge, Mass: Harvard University Press.

— (1983) *Promethean Fire*, Cambridge, Mass: Harvard University Press.

— (1985) 'The relation between biological and cultural evolution', *Journal of Social and Biological Structures* 8: 343-59.

Lyell, C. (1830-3) *The Principles of Geology*, 1st ed., London: Murray.

Lyell, K. (ed.) (1881) *Life, Letters and Journals of Sir Charles Lyell, Bart.*, London: Murray.

Mackie, J.L. (1977) *Ethics: Inventing Right and Wrong*, Harmondsworth: Penguin Books.

— (1978) 'The law of the jungle', *Philosophy* 53: 553-73.

— (1979) *Hume's Moral Theory*, London: Routledge & Kegan Paul.

Marchant, J. (ed.) (1916) *Alfred Russel Wallace: Life and Reminiscences*, New York: Harper.

Matuyama, M. (1929) 'On the direction of magnetisation of basalt in Japan, Tyosen and Manchuria', *Japanese Academy Proceedings* 5: 203-5, reprinted in A. Cox (ed.) (1973) *Plate Tectonics and Geomagnetic Reversals*, San Francisco: Freeman: 154-6.

Marvin, U.B. (1973) *Continental Drift: The Evolution of a Concept*, Washington, DC: Smithsonian Institution Press.

Maynard Smith, J. (1969) 'The status of neo-Darwinism', in C.H. Waddington (ed.) *Towards a Theoretical Biology*, Edinburgh: University of Edinburgh Press: 22, 82-9.

— (1976) 'Group selection', *Quarterly Review of Biology* 51: 277-83.

— (1978a) 'The evolution of behavior', *Scientific American* 239 (3): 176-93.

— (1978b) *The Evolution of Sex*, Cambridge: Cambridge University Press.

— (1978c) 'Optimization theory in evolution', *Annual Review of Ecology and Systematics* 9: 31-56.

— (1981) 'Did Darwin get it right?', *London Review of Books* 3 (11): 10-11.
— (1982) *Evolution and the Theory of Games,* Cambridge: Cambridge University Press.
— (1983) 'Current controversies in evolutionary biology', in M. Greene (ed.) *Dimensions of Darwinism,* Cambridge: Cambridge University Press.
Mayr, E. (1942) *Systematics and the Origin of Species,* New York: Columbia University Press.
— (ed.) (1957) *The Species Problem,* Washington, DC: American Association for the Advancement of Science 50.
— (1963) *Animal Species and Evolution,* Cambridge, Mass: Harvard University Press.
— (1969) *Principles of Systematic Zoology,* New York: McGraw-Hill.
— (1972a) 'The nature of the Darwinian revolution', *Science* 176: 981-89.
— (1972b) 'Sexual selection and natural selection', in B. Campbell (ed.) *Sexual Selection and the Descent of Man 1871-1971,* Chicago: Aldine.
— (1974) 'Teleological and teleonomic: a new analysis', in R. Cohen (ed.) *Boston Studies in the Philosophy of Science* 14: 91-117.
— (1976) 'Is the species a class or an individual?', *Systematic Zoology* 25: 192.
— (1982) *The Growth of Biological Thought,* Cambridge, Mass.: Harvard University Press.
McKinney, H.L. (1972) *Wallace and Natural Selection,* New Haven: Yale University Press.
McMullin, E. (1980) 'Persons in the universe', *Zygon* 15: 69-89.
— (1983) 'Values in science', in P.D. Asquith and T. Nickles (eds) *PSA 1982,* East Lansing, Mich: Philosophy of Science Association: 2: 3-28.
Meteyard, E. (1871) *A Group of Englishmen 1795-1815,* London: Longmans, Green.
Mill, J.S. (1843) *System of Logic,* London: Longman.
— (1910) *Utilitarianism,* London: Dent. First published 1863.
Mishler, B.D. and Donoghue, M.J. (1982) 'Species concepts: a case for pluralism', *Systematic Zoology* 31: 503-11.
Montague, A. (ed.) (1980) *Sociobiology Examined,* Oxford: Oxford University Press.
— (1984) *Science and Creationism,* Oxford: Oxford University Press.
Moore, G.E. (1903) *Principia Ethica,* Cambridge: Cambridge University Press.
Morgan, E. (1973) *The Descent of Woman,* New York: Bantam Books.
Murphy, J. (1982) *Evolution, Morality, and the Meaning of Life,* Totowa, NJ: Rowman & Littlefield.
Nagel, E. (1961) *The Structure of Science,* London: Routledge & Kegan Paul.
— (1977) 'Teleology revisited', *Journal of Philosophy* 74: 261-301.

Nozick, R. (1981) *Philosophical Explanations*, Cambridge, Mass: Harvard University Press.

Ospovat, D. (1981) *The Development of Darwin's Theory*, Cambridge: Cambridge University Press.

Oster, G.F. and Wilson, E.O. (1978) *Caste and Ecology in the Social Insects*, Princeton: Princeton University Press.

Owen, R. (1834) 'On the generation of the marsupial animals, with a description of the impregnated uterus of the kangaroo', *Philosophical Transactions*: 333-64.

— (1849) *On the Nature of Limbs*, London: Voorst.

Paley, W. (1802) *Natural Theology*, in *Collected Works*, vol.4, (1819) London: Rivington.

Pilbeam, D. (1984) 'The descent of Hominoids and Hominids', *Scientific American* 250(3): 84-97.

Plafker, G. (1965) 'Tectonic deformation associated with the 1964 Alaska Earthquake', *Science* 148: 1675-87.

Popper, K.R. (1972) *Objective Knowledge*, Oxford: Oxford University Press.

— (1974) 'Darwinism as a metaphysical research programme', in P.A. Schilpp (ed.) *The Philosophy of Karl Popper*, LaSalle, Ill: Open Court: 133-43.

Puccetti, R. (1968) *Persons: A Study of Possible Moral Agents in the Universe*, London: Macmillan.

Putnam, H. (1975) 'The meaning of meaning', in H. Putnam, *Mind, Language, and Reality*, Cambridge: Cambridge University Press: 215-71.

— (1981) *Reason, Truth and History*, Cambridge: Cambridge University Press.

Quine, W.V.O. (1969) 'Natural kinds', in W.V.O. Quine, *Ontological Relativity and Other Essays*, New York: Columbia University Press: 114-38.

Rawls, J. (1971) *A Theory of Justice*, Cambridge, Mass: Harvard University Press.

— (1980) 'Kantian constructivism in moral theory', *Journal of Philosophy* 77: 515-72.

Raup, D.M. and Stanley, S.M. (1978) *Principles of Paleontology*, 2nd ed., San Francisco: Freeman.

Reed, E. (1975) *Woman's Evolution*, New York: Pathfinder Press.

Reif, W.E. (1983) 'Evolutionary theory in German paleontology', in M. Grene (ed.) *Dimensions of Darwinism*, Cambridge: Cambridge University Press: 173-203.

Romanes, E.R. (ed.) (1895) *Life and Letters of George John Romanes*, London: Longman.

Rosenberg, A. (1980) *Sociobiology and the Preemption of Social Science*, Baltimore: Johns Hopkins University Press.

Rudwick, M.J.S. (1969) 'The strategy of Lyell's *Principles of Geology*', *Isis* 61: 5-33.

— (1972) *The Meaning of Fossils*, London: Macdonald.

— (1974) 'Darwin and Glen Roy: a "great failure" in scientific method?', *Studies in History and Philosophy of Science* 5: 97-185.
— (1976) 'Charles Lyell speaks in the lecture theatre', *British Journal for the History of Science* 32: 147-55.
Ruse, M. (1971) 'Natural selection in *The Origin of Species*', *Studies in History and Philosophy of Science* 1: 311-51.
— (1973a) 'The nature of scientific models: formal v material analogy', *Philosophy of the Social Sciences* 3: 63-80.
— (1973b) 'The value of analogical models in science', *Dialogue* 12: 246-53.
— (1973c) *The Philosophy of Biology*, London: Hutchinson.
— (1975a) 'The relationship between science and religion in Britain, 1830-1870', *Church History* 44: 505-22.
— (1975b) 'Darwin's debt to philosophy: an examination of the influence of the philosophical ideas of John F. W. Herschel and William Whewell on the development of Charles Darwin's theory of evolution', *Studies in History and Philosophy of Science* 6: 159-81. (Also this volume.)
— (1975c) 'Charles Darwin's theory of evolution: an analysis', *Journal of the History of Biology* 8: 60-79.
— (1975d) 'Charles Darwin and artificial selection', *Journal of the History of Ideas* 36: 339-50.
— (1976a) 'Charles Lyell and the philosophers of science', *British Journal for the History of Science* 9: 121-31.
— (1976b) 'The scientific methodology of William Whewell', *Centaurus* 20: 227-57.
— (1976c) 'Reduction in genetics', in R.S. Cohen and A.C. Michalos (eds) *PSA 1974*, Dordrecht: Reidel: 633-52.
— (1977a) 'Is biology different from physics?', in R. Colodny (ed.) *Laws, Logic, Life, Pittsburgh Studies in The Philosophy of Science*, Pittsburgh: Pittsburgh University Press.
— (1977b) 'Karl Popper's philosophy of biology', *Philosophy of Science* 44: 638-61.
— (1977c) 'Sociobiology: sound science or muddled metaphysics?', in F. Suppe and P. Asquith (eds) *PSA 1976*, East Lansing, Mich: Philosophy of Science Association: 2, 48-73.
— (1978) 'Sociobiology: A philosophical analysis', in A. Caplan (ed.) *The Sociobiology Debate*, New York: Harper & Row.
— (1979a) *The Darwinian Revolution: Science Red in Tooth and Claw*, Chicago: University of Chicago Press.
— (1979b) *Sociobiology: Sense or Nonsense?*, Dordrecht: Reidel.
— (1980a) 'Charles Darwin and group selection', *Annals of Science* 37: 615-30. (Also this volume.)
— (1980b) 'Ought philosophers consider scientific discovery? A Darwinian case study', in T. Nickles (ed.) *Scientific Discovery: Case Studies*, Dordrecht: Reidel: 131-50.
— (1980c) 'Philosophical aspects of the Darwinian Revolution', in Summer, L.W., Slater, J.G., and Wilson, F. (eds) *Pragmatism and*

Purpose, Toronto: University of Toronto Press: 220-35.
— (1981a) 'Darwin's theory: an exercise in science', *New Scientist* 90: 828-30.
— (1981b) 'Teleology redux', in J. Agassi (ed.) *Scientific Philosophy Today*, Dordrecht: Reidel: 299-310.
— (1981c) *Is Science Sexist? and Other Problems in the Biomedical Sciences*, Dordrecht: Reidel.
— (1981d) 'Is human sociobiology a new paradigm?', *Philosophical Forum* 13: 119-43.
— (1981e) 'What kind of revolution occurred in geology?', in P. Asquith and I. Hacking (eds) *PSA 1978*, East Lansing, Mich: Philosophy of Science Association: 2, 240-73. (Also this volume.)
— (1981f) 'Are there gay genes? Sociobiology and homosexuality', *Journal of Homosexuality* 6(4): 5-34.
— (1982a) Review of S.J. Gould, *The Mismeasure of Man*, *Isis* 73: 430-1.
— (1982b) 'Darwin's legacy', in R. Chapman (ed.) *Charles Darwin: A Centennial Commemorative*, Wellington, NZ: Nova Pacifica: 323-54.
— (1982c) *Darwinism Defended: A Guide to the Evolution Controversies*, Reading, Mass: Addison-Wesley.
— (1983) 'The new dualism: "*res philosophica*" and "*res historica*"', in M. Ruse (ed.) *Nature Animated*, Dordrecht: Reidel: 3-25.
— (1984) 'The morality of the gene', *The Monist* 67: 167-99.
— (1985) *Sociobiology: Sense or Nonsense?*, 2nd ed., Dordrecht: Reidel.
— (1986a) 'Evolutionary ethics: a phoenix arisen', *Zygon* 21: 95-112.
— (1986b) *Taking Darwin Seriously: A Naturalistic Approach to Philosophy*, Oxford: Basil Blackwell.
— (1987a) 'Is sociobiology a new paradigm?', *Philosophy of Science* 54: 98-104.
— (1987b) 'Darwin and determinism', *Zygon* 22: 419-42.
— (1988a) *Homosexuality: A Philosophical Inquiry*, Oxford: Basil Blackwell.
— (ed.) (1988b) *But Is It Science? The Philosophical Question in the Creation/Evolution Controversy*, Buffalo: Prometheus.
— and Wilson, E.O. (1985) 'The evolution of ethics', *New Scientist* 1478: 108-28.
— and —(1986) 'Moral philosophy as applied science', *Philosophy* 61: 173-92.
Russell, E.S. (1916) *Form and Function*, London: Murray.
Sahlins, M. (1976) *The Use and Abuse of Biology*, Ann Arbor: University of Michigan Press.
Schaffner, K.F. (1967) 'Approaches to reduction', *Philosophy of Science* 34: 137-47.
— (1969) 'The Watson-Crick model and reductionism', *British Journal for the Philosophy of Science* 20: 325-48.
— (1976) 'Reductionism in biology: prospects and problems', in R.S. Cohen and A.C. Michalos (eds) *PSA 1974*, Dordrecht: Reidel: 613-32.
Schlesinger, G. (1963) *Method in the Physical Sciences*, New York: The

Humanities Press.
Schopf, J.W. (1978) 'The evolution of the earliest cells', *Scientific American*: 110-38.
Sebright, J. (1809) 'The art of improving the breeds of domestic animals', in a letter addressed to the Right Hon. Sir Joseph Banks, K.B., London.
Seemanová, E. (1971) 'A study of children of incestuous matings', *Human Heredity* 21(1): 108-28.
Segerstrale, U. (1986) 'Colleagues in conflict: an "in vivo" analysis of the sociobiology controversy', *Biology and Philosophy* 1: 53-88.
Shepher, J. (1979) *Incest: The Biosocial View*, Cambridge, Mass: Harvard University Press.
Simpson, G.G. (1944) *Tempo and Mode in Evolution*, New York: Columbia University Press.
— (1961) *Principles of Animal Taxonomy*, New York: Columbia University Press.
Singer, P. (1975) *Animal Liberation*, New York: New York Review.
— (1981) *The Expanding Circle: Ethics and Sociobiology*, New York: Farrar, Straus, & Giroux.
Smith, R. (1972) 'Alfred Russel Wallace: philosophy of nature and man', *British Journal for the History of Science* 6: 177-99.
Sober, E. (1980) 'Evolution, population thinking and essentialism', *Philosophy of Science* 47: 350-83.
— (1984) *The Nature of Selection: Evolutionary Theory in Philosophical Focus*, Cambridge, Mass: MIT Press.
Sokal, R.R. (1973) 'The species problem reconsidered', *Systematic Zoology* 22: 360-74.
Sommerhoff, G. (1950) *Analytical Biology*, Oxford: Oxford University Press.
Spencer, H. (1892) *Principles of Ethics*, London: Williams & Norgate.
Splitter, L. (1982) 'Natural Kinds and Biological Species', unpublished Oxford University D.Phil thesis.
Stanley, S.M. (1979) *Macroevolution: Pattern and Process*, San Francisco: W.H. Freeman.
Stauffer, R.C. (1959) 'On the Origin of Species: an unpublished version', *Science* 130: 1449-52.
Stebbins, G.L. and Ayala, F.J. (1981) 'Is a new evolutionary synthesis necessary?', *Science* 213: 967-71.
Stidd, B.M. (1980) 'The neotenous origin of the pollen organ of the gymnosperm *Cycadeoidea* and the implications for the origin of higher taxa', *Paleobiology* 6: 161-7.
Strong, E.W. (1955) 'William Whewell and John Stuart Mill: their controversy about scientific knowledge', *Journal of the History of Ideas* 16: 209-31.
Suppe, F. (ed.) (1974) *The Structure of Scientific Theories*, Urbana: University of Illinois Press.
Sykes, L.R. (1967) 'Mechanism of earthquakes and the nature of faulting on the mid-ocean ridges', *Journal of Geophysical Research*

72: 2131-51, reprinted in A. Cox (ed.) (1973) *Plate Tectonics and Geomagnetic Reversals*, San Francisco, Freeman: 332-57.

Symons, D. (1979) *The Evolution of Human Sexuality*, New York: Oxford University Press.

Taylor, P.W. (ed.) (1978) *Problems of Moral Philosophy*, Belmont, Calif: Wadsworth.

Thompson, P. (1983) 'Tempo and mode in evolution: punctuated equilibria and the modern synthetic theory', *Philosophy of Science* 50: 432-52.

Todhunter, I. (1976) *William Whewell D.D., An Account of His Writings with Selections from his Literary and Scientific Correspondence*, London: Macmillan.

Toulmin, S. (1972) *Human Understanding*, Oxford: Oxford University Press.

Trigg, R. (1982) *The Shaping of Man*, Oxford: Basil Blackwell.

Trivers, R.L. (1971) 'The evolution of reciprocal altruism', *Quarterly Review of Biology* 46: 35-57.

—— (1972) 'Parental investment and sexual selection', in B. Campbell (ed.) *Sexual Selection and the Descent of Man, 1871-1971*, Chicago: Aldine.

—— (1974) 'Parent-offspring conflict', *American Zoologist* 14: 249-64.

—— (1976) Foreword to R. Dawkins, *The Selfish Gene*, Oxford: Oxford University Press: v-vii.

—— (1983) 'The evolution of sex', *Quarterly Review of Biology* 58: 62-7.

—— (1985) *Social Evolution*, Menlo Park: Benjamin-Cummings.

—— and Hare, H. (1976) 'Haplodiploidy and the evolution of social insects', *Science* 191: 249-63.

Turner, J.R.G. (1983) '"The hypothesis that explains mimetic resemblance explains evolution": the gradualist-saltationist schism', in M. Grene (ed.) *Dimensions of Darwinism*, Cambridge: Cambridge University Press: 129-69.

van den Berghe, P. (1979) *Human Family Systems: An Evolutionary View*, New York: Elsevier.

—— (1983) 'Human inbreeding avoidance: culture in nature', *Behavioral and Brain Sciences* 6: 91-124.

Valentine, J.W. (1978) 'The evolution of multicellular plants and animals', *Scientific American* 235(3): 140-58.

Van Valen, L. (1976) 'Ecological species, multispecies, and oaks', *Taxon* 25: 233-9.

Vine, F.J. and Matthews, D.M. (1963) 'Magnetic anomalies over oceanic ridges', *Nature* 199: 947-9, reprinted in A. Cox (ed.) (1973) *Plate Tectonics and Geomagnetic Reversals*, San Francisco: Freeman: 232-7.

Vorzimmer, P.J. (1970) *Charles Darwin: The Years of Controversy*, Philadelphia: Temple University Press.

Waddington, C.H. (1957) *The Strategy of the Genes*, London: Allen & Unwin.

—— (ed.) (1969) *Towards a Theoretical Biology*, 2, *Sketches*. Edinburgh:

Edinburgh University Press.

Wade, M.J. (1978) 'A critical review of the models of group selection', *Quarterly Review of Biology* 53: 101-14.

Wade, N. (1979) 'Supreme Court to say if life is patentable', *Science* 206: 664.

— (1980a) 'Supreme Court hears argument on patenting life forms', *Science* 208: 31-2.

— (1980b) 'Court says lab-made life can be patented', *Science* 208: 1445.

Wake, D. (1980) 'A view of evolution', *Science* 210: 1239-40.

Wallace, A.R. (1855) 'On the law which has regulated the introduction of new species', *Annals and Magazine of Natural History* 16: 184-96.

— (1858) 'On the tendency of varieties to depart indefinitely from the original type', *Journal of the Proceedings of the Linnean Society, Zoology* 3: 53-62.

— (1869) 'Sir Charles Lyell on geological climates and the origin of species', *Quarterly Review* 126: 359-94.

— (1870) 'The limits of natural selection as applied to man', in A.R. Wallace, *Contributions to the Theory of Natural Selection*, London: Macmillan.

— (1889) *Darwinism*, London: Macmillan.

— (1900) *Studies: Scientific and Social*, London: Macmillan.

— (1905) *My Life: A Record of Events and Opinions*, London: Chapman & Hall.

Wallwork, E. (1982) 'Thou shalt love thy neighbour as thyself: the Freudian critique', *Journal of Religious Ethics* 10: 264-319.

Washburn, S.L. (1951) 'The new physical anthropology', *Transactions of the New York Academy of Science* 2(13): 298-304.

Wegener, A. (1915) *Die Enstehung der Kontinente und Ozeane*, Braunschweig: Vieweg.

Weinrich, J.D. (1976) 'Human Reproductive Strategy, I: Environmental Predictability and Reproductive Strategy; Effects of Social Class and Race, II: Homosexuality and Non-Reproduction; Some Evolutionary Models', unpublished Ph.D. thesis, Harvard University.

— (1978) 'Is homosexuality biologically unnatural?', unpublished manuscript.

— (1982) 'Is homosexuality biologically natural?', in W. Paul, J.D. Weinrich, J.C. Gonsiorek, and M.E. Motvedt (eds) *Homosexuality: Social, Psychological, and Biological Issues*, Beverly Hills: Sage: 197-208.

West Eberhard, M.J. (1975) 'The evolution of social behavior by kin selection', *Quarterly Review of Biology* 50: 1-33.

Westermark, E. (1891) *The History of Human Marriage*, London: Macmillan.

Whewell, W. (1832) 'Principles of Geology ... by Charles Lyell ... vol. II', *Quarterly Review* 47: 103-32.

— (1833a) Address to the British Association, *Proceedings of the*

British Association for the Advancement of Science, 1833.

— (1833b) *Astronomy and General Physics Considered with Reference to Natural Theology: Bridgewater Treatise III,* London: William Pickering, 7th ed., 1863.

— (1837) *The History of the Inductive Sciences,* London: Parker.

— (1840) *The Philosophy of the Inductive Sciences,* London: Parker.

— (1853) *Of the Plurality of Worlds: An Essay,* London: Parker.

White, M.J.D. (1978) *Modes of Speciation,* San Francisco: Freeman.

Whitehouse, M.L.K. (1959) 'Cross- and self-fertilization in plants', in P.R. Bell (ed.) *Darwin's Biological Work,* Cambridge: Cambridge University Press: 207-61.

Wickler, W. (1973) *The Sexual Code: The Social Behavior of Animals and Men,* Garden City: Doubleday.

Wiggins, D. (1973) *Sameness and Substance,* Oxford: Basil Blackwell.

Wiley, E. (1978) 'The evolutionary concept reconsidered', *Systematic Zoology* 27: 17-26.

— (1980) 'Is the evolutionary species fiction? – A consideration of classes, individuals, and historical entities', *Systematic Zoology* 29: 76-80.

— (1981) *Phylogenetics: The Theory and Practice of Phylogenetic Systematics,* New York: John Wiley.

Wilkinson, J. (1820) Remarks on the improvement of cattle, etc., in a letter to Sir John Saunders Sebright, Bart. MP, Nottingham.

Williams, G.C. (1966) *Adaptation and Natural Selection: A Critique of Some Current Evolutionary Thought,* Princeton: Princeton University Press.

— (1975) *Sex and Evolution,* Princeton: Princeton University Press.

Williamson, P. (1981) 'Palaeontological documentation of speciation in cenozoic molluscs from the Turkana Basin', *Nature* 293: 437-43.

Wilson, E.O. (1975a) *Sociobiology: The New Synthesis,* Cambridge, Mass: Harvard University Press.

— (1975b) 'Human decency is animal', *The New York Times Magazine* 12: 38-50.

— (1978) *On Human Nature,* Cambridge, Mass: Harvard University Press.

Wilson, J.T. (ed.) (1970) *Continents Adrift,* San Francisco: Freeman.

Wilson, L. (ed.) (1970) *Sir Charles Lyell's Scientific Journals on the Species Question,* New Haven: Yale University Press.

— (1972) *Charles Lyell: The Years to 1841: The Revolution in Geology.* New Haven: Yale University Press.

Wynne-Edwards, V.C. (1962) *Animal Dispersion in Relation to Social Behaviour,* Edinburgh: Oliver & Boyd.

INDEX